INTELLIGENCE IN THE FLESH

INTELLIGENCE IN THE FLESH

WHY YOUR MIND NEEDS YOUR BODY MUCH MORE THAN IT THINKS

GUY CLAXTON

YALE UNIVERSITY PRESS
NEW HAVEN AND LONDON

For information about this and other Yale University Press publications, please contact:
U.S. Office: sales.press@yale.edu   www.yalebooks.com
Europe Office: sales@yaleup.co.uk   www.yalebooks.co.uk

Typeset in Minion Pro by IDSUK (DataConnection) Ltd
Printed in Great Britain by TJ International Ltd, Padstow, Cornwall

Library of Congress Control Number: 2015940459

ISBN 978-0-300-20882-5

A catalogue record for this book is available from the British Library.

10 9 8 7 6 5 4 3 2 1

MIX
Paper from
responsible sources
FSC
www.fsc.org   FSC® C013056

We only believe in those thoughts which have been conceived not in the brain but in the whole body.

W. B. Yeats, *Essays and Introductions*, Palgrave Macmillan, 1961, p. 235

To Jude, with love

CONTENTS

FIGURES

LIMBERING UP

AN INTRODUCTION

If the body had been easier to understand, nobody would have thought that we had a mind.

Richard Rorty[1]

Over the last century, human beings in affluent societies have become more and more sluggish. Millions of us work in offices, pushing paper, staring at screens, discussing proposals and re-arranging words and spreadsheets. For our leisure, we look at more screens, text and tweet, escape into virtual worlds, gossip and chatter. Some of us still play tennis or knit, but the drift is undeniably chair- and couch-wards. Our functional bodies have shrunk: just ears and eyes on the input side, and mouths and fingertips on the output side. Laundry now involves all the physical skill and effort of pushing clothes through a porthole and pushing a button. Cooking can be no more than ripping off a plastic film and closing the microwave door. Our real bodies get so little attention, and so little skilful use, that we have to make special arrangements to remember them: we program country walks and trips to the gym into our smartphones.

Inactivity and cleanliness used to be the privilege of the rich: not any more. And the machines that make all this leisure possible are opaque – most of us wouldn't know how to fix them, and wouldn't want to. We have become mind rich and body poor.[2]

But this is not another panicky book about obesity, heart disease or the dangers of the internet. Nor is it a hymn of nostalgia for the dying arts of quilting and whittling. At the heart of this book is an argument: that we neglect our bodies because we underestimate their intelligence. The problem is not that we have become 'lazy', or devoid of 'willpower'. It is a matter of assumptions and values. We aspire to cerebral jobs and disembodied pastimes because we have got the idea that those kinds of things require more *intelligence* than practical, physical things, and consequently they are more highly esteemed in our societies. Crudely, they make us look *smarter*, and looking smart is good, so doing mind stuff makes us feel good. (Of course, because they are more highly esteemed, they also, by and large, pay more.) Conversely (with a few possible exceptions such as some top athletes) being physically tired, dirty and smelly is associated with a lack of intelligence. So we learn to aspire to being clean and verbal.[3]

We still think about the relative intelligence of body and mind in an archaic and inaccurate way: so says the new science of embodied cognition. Many neuroscientists do not now think that intelligence belongs only to minds, and that the pinnacle of human intelligence is rational argument. They no longer believe that the mind is an ethereal source of control, sent to curb the body's waywardness and compensate for its stupidity. They do not think that minds and bodies are different kinds of stuff. The idea that bodies are dumb vehicles and minds are smart drivers is old hat. The new science of embodiment has important implications for how we think about ourselves and how we live our lives. This book is a shot at getting that knowledge out there – because I think it matters a lot.

The predominant association of intelligence with thinking and reasoning isn't fact; it is a cultural belief – a virulent meme, some would call it – that misdirects us. Young people who prefer doing intricate things with their bodies – breakdancing, skateboarding – to doing their maths homework are not lacking in intelligence. I think they are part of a growing cultural rebellion against the hegemony of the intellect (though most of them wouldn't put it quite like that). I hope this book will help their parents and teachers understand why that rebellion is itself intelligent. I hope it might contribute to a wider revaluing of the practical and physical, in education for example, so that those who are not cerebrally inclined will not be led to make the mistake of feeling stupid.

*****

Let me, in this overture, introduce some of the main themes that will emerge as the scientific story unfolds.

The recurring motif is this: we do not *have* bodies; we *are* bodies. If my body was different, I would *be* different. If I was made of silicon or fibre optics, I would need different things, respond to different things, notice different things, and be intelligent in a different kind of way. My mind was not parachuted in to save and supervise some otherwise helpless concoction of dumb meat. No, it's just the other way round: my intelligent flesh has evolved, as part of its intelligence, strategies and capacities that I think of as my 'mind'. I am smart precisely because I am a body. I don't own it or inhabit it; from it, I arise.

This realisation is both completely mundane – and quite extraordinary. It overturns the accepted, intuitive psychology –

academics call it the 'folk psychology' – of two thousand years of Western civilisation. Chapter 2 sets the scene for the new view by taking a quick scan back over the evolution of the old view. From classical Greece to the late twentieth century, it was simply inconceivable that a pillar of meat – and especially the dull-looking lump of matter inside the skull – could have been the source of Euclid's geometrical proofs, Plato's *Republic* or Shakespeare's sonnets; or that acts of great selflessness and wise judgement could have arisen *sui generis* from 70 kilos or so of flesh. The smart stuff had, then, to be immaterial and come from elsewhere. The 'mind' was invented to fill what philosophers call 'the explanatory gap'. Consciousness, especially rational thinking, looked to our forebears as if it had to sit in the centre of this hypothetical mind, with the Senses delivering information to it through one bodily door, and Decisions being despatched to the workhorses of the body through an opposite one. We think that we See things, then we Think about them, then we make Decisions, and finally we Act. But it's not like that at all.

Chapters 3, 4 and 5 take us into the modern, scientific understanding of the body. When science first tried to 'naturalise' the mind, its most obvious physical accomplice was the brain. But, as I will show, the proper substrate of the mind is not the brain alone but the entire body. I'll unfold a view of the human body as a massive, seething, streaming collection of interconnected communication systems that bind the muscles, the stomach, the heart, the senses and the brain so tightly together that no part – especially the brain – can be seen as functionally separate from, or senior to, any other part. Torrents of electrical and chemical messages are continually coursing throughout the entire body and its brain. In fractions of a second, the 'decision-making' of the brain can be influenced by a badly behaved bacterium in the gut, and the level of sugar in the blood can be altered by a squeak or a dream. The cells and molecules of the immune system have so many receptors at all levels in the brain that the immune system now has to be thought of as an integral part of the central nervous system. In fact, it's all just one system.

I'll demonstrate that we are fundamentally built for action, not for thinking or understanding, and that, as a consequence, our intelligence is deeply orientated towards the construction of effective and appropriate behaviour. Thinking is a recently evolved tool for supporting smart action. We'll see that the brain evolved to help increasingly complicated bodies coordinate their interlocking sub-systems in the service of the whole community. Brain is servant, not master of the body. It's a chatroom, not a directorate. Seeing, Thinking, Deciding and Acting are not strung out, like different departments in a factory; they are inextricably entwined. Careful science shows that how I see is instantly imbued with what I want and how I might act. The body-brain is designed to blend all these influences together in the blink of an eye, and often issues intricate, intelligent actions without thought or premeditation.

This being so, we need to rethink the relationship between thoughts and feelings. Feelings are not a nuisance. They are not – as Plato thought, and many still do – wayward and primitive urgings that continually threaten to undermine the fragile structures built by dispassionate reason. They are, as we will see in Chapter 6, the bodily glue that sticks our reasoning and our common sense together. Feelings are somatic events that embody our values and concerns. They signal what we care about: what gives our lives meaning and direction. Our hopes and fears arise from the resonance of our organs in response to events. Without physical feelings and intuitions, abstract

5

intelligence sheers away from the subtleties and complexities of the real world, and people become 'clever-stupid', able to explain and comprehend but incapable of linking that understanding to the needs and pressures of everyday life. Particular emotions can get tangled and perverted by experience, and very often do. We become fearful of intimacy, or angry at our own timidity. But the solution is not for Reason to trounce Emotion. The body's signals are essentially wise, if sometimes confused. If the wisdom is ignored, it will be hard to sort out the confusion.

Language and reason themselves look different when we see that they too are rooted in the body. Chapter 7 explores the ways in which our more abstract understanding grows out of the physical and sensory concepts that the young child grasps first – giving and taking, coming and going, full and empty, warm and cool, nurturing and threatening. We come to under-stand what someone means when they ask if we have *grasped* the argument, by analogy with the physical act of grasping. And these primeval links back to the body are never lost. There is no separate bit of the brain where abstract ideas like Truth and Justice are stored, and where Philosophising takes place. From birth till death, the body is the moment-to-moment substrate of our thoughts and desires – however refined. Studies show that, in complicated predicaments, people make better deci-sions when they rely on their 'gut feelings' as well as their reason and do not see them as antagonists.

Language itself teems with expressions that muddle up mind and body. I hear a side-splitting joke and laugh till I cry or hoot with glee (or the joke may be lame, toe-curlingly awful, eliciting only a weak groan and a rolling of the eyes). I read a heart-rending story and am moved to (a different kind of) tears. My eyes pop with surprise and I prick up my ears. My shoulders slump with disappointment, I have butterflies in my stomach, and my blood runs cold. I feel gripped by an idea, or queasy at the very thought. I know things in my bones and feel them in my water. A creepy tale makes my flesh crawl, while a compliment makes me flush with pride. A memory comes and I smile quietly to myself. Informally, instinctively, we know that mental and bodily events resonate tightly with each other – but all these somatic reactions are no mere accoutrements of the mental activities they accompany; they are absolutely of the essence.

Much of our somatic intelligence operates unconsciously, without conscious supervision or even awareness. So what is consciousness for, and how does it emerge from the intrinsic activities of a complicated body? In Chapter 8 I'll suggest that conscious thoughts and images are actually the result of a progressive (though often quite rapid) process of unfurling meanings and decisions that have their origins in the darker, deeper, more visceral areas of the brain and body. Thoughts are stories the embodied brain constructs about what is going on in its own hidden depths: reports from the interior, sometimes heavily edited and censored, and sometimes arriving by pigeon post long after the action is over. Many experiments show that our conscious intellect is often a rather pale reflection, or even a crude caricature, of the sophisticated operations that are going on 'behind the scenes'. Consciousness has its own priorities – creating a semblance of order and self-esteem, for example – which lead it, often, to misrepresent the complexity and waywardness of what is going on below. We confabulate much more than we like to think.[4]

Bodies do not stop at the skin. So neither do minds. We'll see in Chapter 9 that the internal streaming of information continues through our fingertips and out into the tools we use, for example. When you pick up a familiar tool, be it a fish slice or a chisel, your brain literally incorporates it into its representation of your body;

it becomes as much a part of your body as the hand itself. It is easy to trick your body-brain into believing that a rubber arm on a table in front of you is actually your own, so that, when someone hits it with a hammer, you cannot but flinch. But it's more even than that. We are also deeply interconnected, through our bodies, largely unconsciously, with the material and social worlds around us – our bodies literally reverberate with each other at many levels. The 'intelligent agent', seen rightly, extends throughout and beyond the whole body. It is constituted by the tools and the space around us, and also by everyone with whom we are 'in touch'.

The fact that we are fundamentally doers means we are also inveterate makers. Making is doing that involves those extraordinarily sophisticated on-board tools, our hands. In Chapter 10 we will find that human intelligence lives in our hands just as much as in our tongues and our brains. Making is in our blood, it seems. We have been crafted by evolution to be natural-born engineers, compulsive sculptors of our environments. Human beings are habitat decorators, toolmakers and workshop designers par excellence; we were *Homo fabricans* long before we were *sapiens*. Or rather, the *sapiens* grew out of the *fabricans*, and still relies deeply upon it. It is in our nature to amplify our intelligence by imagining, and then making, ever more powerful tools. We are only as smart as we are because we are enmeshed in a world of our own making: a vast web of books, spectacles, notes, printers, weblinks, diaries, calendars, maps, satellite navigation gizmos, computer programs, filing systems, Skype links, mobile telephones . . . all of which I know, more or less, how to capitalise on. As Andy Clark puts it, 'we make our worlds smart so we can be dumb in peace'.[5] My intelligence stretches way beyond what can be captured in an IQ test.

There are signs of a wider resurgence of the physical: a backlash, perhaps, against the intellectualisation of intelligence.

Optimistically, you could discern evidence of a New Materialism on the rise: one which is not about conspicuous consumption, but about the quiet, protracted hands-on pleasures of making, mending, customising and perfecting physical skills. The Maker Movement in the States gathers strength, and puts pressure on manufacturers to make things mendable again. FabLabs, 'tinkering workshops' and 3D printers are springing up in response to the desire to engage with solid, workable stuff. The more the digital world takes hold, the stronger, for many of us, seems the compensatory desire to get back from the virtual to the real, from the symbolic to the material. And this signals a re-esteeming of physical delicacy, sensibility and creativity (beyond those protected 'Sites of Special Cultural Interest' called sport and art). Craft *is* cognition, people are saying. Doing and thinking are not separate faculties; they are inextricably entwined.

So, with all this in mind (and body), we will come back to the question: what does it really mean to be intelligent? A lot has been written in the last twenty years about different kinds of intelligence. We have had emotional intelligence, practical intelligence and 'bodily-kinaesthetic' intelligence, along with a host of others. But I'm not proposing another *kind* of intelligence to add to the list. My contention here is more radical than that. It is that practical, embodied intelligence is the deepest, oldest, most fundamental and most important intelligence of the lot; and the others are aspects or outgrowths of this basic, bodily capability. Emotional intelligence is an *aspect* of bodily intelligence. Mathematical intelligence is a *development* of bodily intelligence. There is a world of difference between human intelligence, properly understood, and mere cleverness.

In the real world, intelligence refers to the optimal functioning of the eco-socio-embodied systems that we are. Intelligence isn't a faculty; it is the behaviour of an entire system when it is able to come up with good answers to the perennial question: What's the best thing to do next? Intelligence is reconciling desires, possibilities and capabilities in real time – especially when the situation is complex, novel or unclear. The ability to figure out the next number in the sequence 1, 2, 3, 5, 8, ... is a very poor proxy for your ability to act wisely when you lose your wallet or when you get a great job offer that would mean uprooting the family. At such times you need to be able to analyse the situation, check your values and assumptions, and figure out the consequences of various courses of action. In times of change or challenge you need your reason – but you also need your ability to sense inwardly what is truly in your own best and deepest interests. And a lot of clever people can't do that. I don't think they teach it (yet) at Harvard Business School.

Which obviously leads us on to the question of how you do 'teach' it, and that is the business of Chapter 11. For the body to be at its most intelligent, it has to be properly 'strung'. The different sub-systems need to be able to talk to each other both directly and via the chat-room of the brain. If these circuits are not able, quite literally, to pick up each other's vibes, then sympathetic resonance doesn't happen, the quality of available information is reduced, and the overall harmony of the total system that we are is degraded. This can happen through injury, illness or ageing, and can often be reversed by physical exercise. But we can also lose harmony by dampening our 'interoceptive awareness', and for this visceral intelligence to be rehabilitated, physical exercise needs to be accompanied by efforts to refocus and sharpen our attention. Dance, yoga and t'ai chi all have proven effects on cognitive functions such as decision-making and problem-solving, for example.

Finally in Chapter 12 we will round up the implications of embodiment both for individual well-being and for the nature of the over-intellectualised, somatically impoverished institutions that surround us. The way we think about intelligence is built into the social structures we create – religion, medicine, government, as well as law and education – so a shift in our view of mind has repercussions not only for individual identity but for public life as well. We design law courts and classrooms in which physical movements and reactions are treated as disruptions, subversive of the serious work of the mind – yet some people think better when they are moving. Why do we make children sit still if intelligence benefits (as it does) from physical movements and gestures? Why do we set up adversarial, argumentative forms of governance and jurisprudence if rational sophistical debate is not the only, or even (very often) the best, form of intelligence we possess? How can we give back to emotion and intuition their proper roles as *constituents* of human intelligence, without tipping over into a kind of New-Agey denial of rationality itself? These are questions which I will only just begin to address. Re-establishing a balanced, embodied society will be tricky, to put it mildly, and will take gradual shifts in understanding by us all. I hope *Intelligence in the Flesh* will contribute to those shifts.

*****

In a way, this book is the third (and probably last) in a trilogy which began in 1997 with my book *Hare Brain, Tortoise Mind: Why Intelligence Increases When You Think Less*. There, I was one of the first to argue, on the basis of scientific research, that much human intelligence depends on processes of which we are not – and largely cannot be – aware. I called it 'the undermind'. It is now also widely known as the adaptive or cognitive unconscious. Then in 2005 I wrote a sequel, *The Wayward*

Mind: An Intimate History of the Unconscious, aiming to situate this new 'unconscious' within a wider cultural and historical context. I brought together the kinds of stories that societies since about 4000 BC have created to try and account for mental phenomena that seem to be at odds with 'common sense': hypnosis, hallucination, mental illness and creative inspiration, for example. Perhaps they arise from the external influences of gods, demons and spirits. Or maybe they spring from the 'subconscious', a dark inner maelstrom of whimsy and wildness (as Plato thought, long before Freud). Or was the source simply activity in the matter of which we were made, that just sometimes failed to conform to normal expectations – an excess of 'black bile', maybe? I showed that versions of these three stories recur and compete throughout human history, right up to the present day. I argued that each has its value, and its pitfalls. Stories can be useful even if (or precisely because) they do not refer to objectively verifiable things.

But with the rise of affective neuroscience and embodied cognition we are now able to offer much more robust and compelling versions of the third story. At my most radical, I would now claim that, not only are 'the gods and spirits' non-existent (even though they may still have their uses), but the unconscious is dead too. We may choose to continue using it as a metaphorical or poetic way of talking, but thar ain't no such animal. There are myriad processes in the body that never lead to conscious experience, but there is no real, identifiable place or agent inside us that is a separate source of impetus from consciousness and reason. Like 'the mind', 'the unconscious' is a place-saver, a dummy explanation. It is like a temporary filling in a tooth, put there till something better comes along. And now it has.

*****

12

I should say a little about my style and source material. I have read hundreds of research papers, but a good deal of this primary material is quite technical and even arcane. I have tried to dig out the main points and present them in an accessible and palatable way; to walk a middle path between respecting the rigour and niceties of research and telling a good story. But this means I will inevitably have skated over many points of contention that my more learned academic colleagues rightly consider to be important. The so-called 'hard problem' of consciousness will have to wait for another day, as, very largely, will the relationship between what I have to say here and other important work on what are called Systems One and Two (or, in Daniel Kahneman's terms, fast and slow thinking). Sorry, guys.

I have also glossed over several topics you may reasonably have expected would receive better treatment in such a book. I'll say nothing about the difference between male and female bodies, or the difference between men's and women's relationship to their bodies. Issues of physical and mental health remain largely unexplored, and what used to be called 'psychosomatic' conditions have not got anything like the attention they deserve. Neither have traditional understandings of the body-mind such as those found in many indigenous cultures, notably the traditional Chinese, Indian and Native American. I now suspect that these systems of thought, and others like them, are intricate blends of real insight and hocus-pocus, but it would take a whole book to disentangle one from t'other, and there isn't room here even to begin to make a start. (Can I feel a quartet coming on?)

As well as the primary sources, I must gratefully acknowledge several overviews of the new science of embodied cognition on which I have drawn, sometimes extensively. They include Roy Porter's *Flesh in the Age of Reason*, Evan Thompson's

Mind in Life, Francisco Varela and others' *The Embodied Mind*, Mark Johnson's *The Meaning of the Body*, Andy Clark's *Being There: Putting Brain, Body and World Together Again*, Shaun Gallagher's *How the Body Shapes the Mind* and Mark Rowlands' *The New Science of the Mind*. I have pilfered part of my title from George Lakoff and Mark Johnson's *Philosophy in the Flesh*, and many ideas from Antonio Damasio's *Descartes' Error* and *Self Comes to Mind*. I am acutely aware that, in writing this book, I am standing on the shoulders of giants, and the above authors are some of them. Most of these books are quite technical, however, both scientifically and philosophically, and they often delve into academic disputes of limited interest (and accessibility) to non-specialists or people new to the field.

There are also three excellent recent books on craft and practical intelligence that I want to mention as well – Matthew Crawford's *The Case for Working with Your Hands*, Richard Sennett's *The Craftsman* and Mike Rose's *The Mind at Work*. None of these, though, situates the renewed interest in physical making within the emerging science of embodied cognition. For those who want to dig deeper, any of these books would make excellent reading. However, none of them has – for good or ill – the sweep of mine, attempting to embrace visceral physiology, brain science, the function of emotion, consciousness, craftsmanship and 'handiness', as well as the wider social and personal implications of the new science for our view of human intelligence. I hope the breadth will prove interesting, even if some of the depth has had to be sacrificed.

Well, that's quite enough throat-clearing. Let's get on with it. We'll start with a quick look at how the separation of mind and body, and the privileging of one over the other, came about.

2

A BRIEF HISTORY OF
ANTI-BODIES

It came to pass
That 'Brother Ass'
(As he his Body named)
Unto the Saint
Thus made complaint:
'I am unjustly blamed'.

John Bannister Tabb, 'St Francis'[1]

One central objective of this book is the reunification of mind, brain and body. In order to see how we might do this, it is useful to rehearse quickly how and why they came to be separated. I can be quite brief, because the history of the mind–body split is reasonably well known. But I need to say something here in order to establish the context for what follows.[2]

For 2,500 years, the human body has had a very mixed press. If we start this brief survey of historical attitudes in classical Greece, we might think at first that they had a better-rounded view of the person than we do. Avery Brundage, the long-time president of the International Olympic Committee, for example, often extolled the 'Golden Age' of Greece for celebrating equally physical prowess and mental dexterity. He claimed that 'philosophers, dramatists, poets, sculptors and athletes met on common ground'. He perpetuated the idea that even the great Plato was also an accomplished athlete: folklore has it that in his youth Plato was a champion wrestler.[3]

The truth, however, seems more prosaic and more ambivalent. If he exercised at all, it is much more likely that Plato merely worked out occasionally at his local gym. And such equality of esteem as there was between the body and the intellect seems, around the fifth century BC, to have been overtaken by a more generalised disdain for physicality. The philosopher Xenophanes captured the shift in opinion when he wrote rather peevishly: 'If a man wins victory in wrestling or boxing, he is [still] given a seat of honour at the games . . . yet he is not as worthy as I am. For my wisdom is better than the strength of humans or horses. It is wholly unfair to rank strength above my wisdom.'[4] As today, it had become more commonplace for the 'middle classes' to go to the theatre, and discuss the play over dinner afterwards, rather than to roar for victory at a wrestling match.

With the rise of intellectual philosophising, and a more 'refined' approach to entertainment and aesthetics, the view developed that wisdom and intellect could self-evidently have nothing to do with the mere brutality of bodies (human as well as animal), so they required a faculty of a completely different, non-physical kind, for which Isocrates co-opted the word *psyche*. (Until then, *psyche* had referred to the animating impulse that caused you to draw breath. First *psyche* was the spirit that caused you to 'inspire', and only later the source of divine 'inspiration'.)[5] The influential Greek belief system known as Orphism, said to derive from the legendary poet Orpheus, referred to the body as 'the tomb of the soul' and encouraged people to look forward to their human death as the release of the soul from its entrapment, and as the beginning of true life. A gold-leaf tablet found in Crete instructed people on how to greet a leader of the Underworld, whom they would meet after death. 'Now you have died and now you have come into being, O Thrice-happy One, and this same day.' Why 'thrice-happy'? Another Orphic fragment, this time recovered from Olbia in Sardinia, explains. It simply says, 'Life. Death. Life. Truth.' You lived, then you died, and then you were born again into the true life – leaving the physical body behind like a discarded chrysalis. The very name Orpheus is thought to derive from an earlier Greek word meaning 'discarded or 'abandoned' (hence 'orphan' in our times).

Soon, the body came to be seen not just as the soul's sarcophagus, but as its enemy too. In Aristotle's proposed education system, boys would not be allowed to pursue physical and mental training in the same year, 'since the two kinds of exercise naturally counteract one another, exertion of the body being an impediment to the intellect', and vice versa. By the second century AD, even the renowned physician Galen would try to dissuade young men from becoming athletes, claiming 'they are so lacking in reasoning they don't even know if they have a brain. They cannot think logically at all – they are as mindless as dumb animals.' The idea that physical virtuosity could count as a form of intelligence was rendered laughable. The 'god-like' image of the young adult male, epitomised in the mythical Adonis, beautiful in form and admirable in fitness and expertise, had become, you might say, more of a niche interest. Even the phrase *mens sana in corpore sano*, a healthy mind in a healthy body, coined by Roman satirist Juvenal in around 345 AD, makes no genuine connection between the two kinds of health. It occurs in the context of a discussion of the kinds of foolish things that people pray for, where Juvenal suggests that, if you must pray for something, just 'pray not to get sick, and not to go crazy'. The phrase seems to have lain dormant until unearthed by one John Hulley in 1861 and misused as the motto of the newly formed Liverpool Athletics Club.

Early Christianity adopted these attitudes, denigrating the body and generally seeing it as a source of distraction and waywardness in constant need of taming. St Paul called the body 'sin's instrument', and suggested that Christians should follow his example – 'I bruise my own body and make it know its master' – because only through such discipline could the soul be protected from temptation and thus improved.[6] St Francis famously referred to his body disparagingly as 'Brother Ass', while medieval Christianity has been characterised as a time of 'hatred of the flesh'. Though there were pockets of resistance to these anti-body sentiments, it wasn't until the establishment of the privileged nineteenth-century English public schools that physical and mental exercise were cautiously allowed to coexist again in the form of so-called 'muscular Christianity'. At the same time, a gay admiration of the young athletic male body – the sexually alluring Adonis – resurfaced in the work (and lives) of Oxford intellectuals such as Walter Pater.

The Greek philosophers also instigated the tight association between 'higher', 'better', 'more abstract' and 'eternal'. Bodies were suspect not just because they could be wayward and impulsive, but because they were impermanent (they die) and dependent on the uncontrollable slosh of experience (they become tired, get injured and catch colds). So where were purity and security to be found? According to Plato, they dwelt in a parallel, abstract – what we would now call 'Platonic' – world of everlasting ideas and concepts. Down Here it can get pretty tricky and ugly ... but Up There, in some hard-to-access

Heaven or Paradise, there are Beauty and Truth beyond corruption. So if people wanted to escape this 'vale of tears', they had to work their way Upwards, into the realms of logic, reason and theology. Rationality and sanctity became yoked together. They had better not bother too much with seeing, tasting and touching (and especially not with recreational sex) but should aspire to believe, pray, hope and think. As the great psychologist William James put it, rather more formally, abstract concepts and relationships (as in mathematics) express eternal verities, and hence:

> There arose a tendency, which has lasted all through philosophy [and public life], to contrast the knowledge of universals, [seen] as god-like, dignified and honourable to the knower, with that of particulars and sensibles, as something relatively base which more allies us with the beasts.[7]

As the human head was physically above the genitals, the anus and the feet (at least when standing and praying), so 'heaven' was spatially located above 'hell'. Up became associated with Good, and Higher was Better; Down became associated with misfortune. Cars and relationships don't just break, they break *down*;[8] when we lose money the stock market *falls* and we are *down* on our luck. And the Lowly is simpler, more mundane and more menial than the Higher. Down also becomes associated with impurity – feet get dusty, shit falls to the ground, watch where you're walking – so Up becomes Bright, Light, Pure and Ethereal. Your head, being nearer to heaven, is where your soul lives. Soul and sole inhabit opposite poles. It was the nebulous soul – or, as it sometimes came to be called, the mind – that got the credit for any more refined thoughts or sensibilities. Where soul and mind were distinguished, mind

19

tended to retain the cerebral or 'cognitive' functions of the original *psyche* while the soul took custody of the more spiritual or divine aspects. In the absence of science, such ghostly agents are free to proliferate.

In Europe it is generally thought to have been René Descartes who succeeded in cleaving Mind from Body completely. To him, it must have seemed obvious that 'there is nothing included in the concept of the body that belongs to the mind; and nothing in that of mind that belongs to the body'.[9] The human brain, for example, is not much to look at. Slice the top off someone's head and what you see is a three-pound wrinkly blob of grey-brown spongy meat that looks singularly unimpressive and inert. To anyone looking at a brain before about a hundred years ago, it would have been literally unthinkable that this dull organ could be intelligent, in any real sense. Likewise, the 'housekeeping' processes with which bodies concerned themselves – breathing, digesting, contracting muscles and so on – did not warrant being called 'intelligent'. Feet did not think; tears did not make decisions. Because the investigators of the day had no way of looking at the body and brain as an incredibly intricate, dynamic whole, to talk of embodied intelligence would have made no sense at all.

Of course, the disparagement of the body, and even the recommendation that the flesh be 'mortified', is not a purely European or Christian idea. In many of the world's religions, the body is seen as a hindrance to spiritual or intellectual progress. The Catholics of Opus Dei may, to this day, bind their thighs with painful metal ligatures, but similar practices of self-flagellation can be found amongst Buddhists and Hindus. Hinduism, for example, proclaims very loudly that 'I am not the body', and so do many versions of Buddhism. Buddhists often see the body, especially its internal organs and secretions, as objects of disgust. They may also see the sensual – or even more broadly the sensory – pleasures of the body as dangerous impediments to 'enlightenment'.

Islam, though it has a somewhat less hostile attitude towards the body than Christianity, also seems to have a horror of what goes on inside, and of what emerges from the inside. Not only urine and faeces but blood and semen are viewed as ritually unclean and as demanding purification. Judaism is a partial exception to this tendency. Body and soul are indeed seen as distinct; however, the aim of the spiritual journey is not their final dissociation but their reunification. The body has its own holiness, which explains Judaism's more positive attitude towards sex. In the rabbinic and Kabbalistic traditions, for example, the preferred time for sex is on the Sabbath, when physical and spiritual pleasure can be unified. A medieval Jewish essay on sex called 'The Holy Letter' explicitly argued against any philosophical or religious attitude that denigrates the sense of touch. It claimed that sex is holy because it 'expresses and enacts the secret of oneness'.[10]

One of the reasons why it is easy to underestimate the body is that it is so secretive. Bodies do not reveal themselves much to the observing eye (or 'I') of conscious awareness – largely because eyes are the very tools whereby that awareness comes about. If you put your ear to a brain (or an ear) you hear no sounds. If you cut open an eye or its associated visual cortex you find no movie show going on inside. Bodies are very largely dark and silent. Eyes do not see themselves – their 'processes' are dissolved in the act of seeing *other* things. We cannot hear our ears hearing (unless perhaps we are suffering from tinnitus); we can only hear the sounds that the ear is making available.

Conscious awareness has inherent limitations, too. For example it cannot spot things that change very fast. We do not notice the flickering of the movie, or of the fluorescent light bulb. And most of our inner workings shimmer too fast for the eye to see. Our eyes make jumps several times a second, but it would not serve us well to be aware of them. We want to 'see' a pretty stable backdrop against which significant changes in the world itself will stand out. We are not built to notice changes in the instrumentation. We watch the movie, not the projector. Contrary to Descartes, we are not even good at observing our own thoughts; most of the time we get so bound up in them that we leap on their backs and gallop away and are incapable of standing back and watching the horse. Even when our body-brains do make their *products* known to consciousness, their *processes* remain silent and invisible. So it is no wonder that we have underestimated our bodies. It is not that bodies are simple – far from it. It is just that it is extremely hard to notice their complexity. For that, we need science.

The mind

So what is this mysterious entity, the mind (or soul), that has to be imported to do the cognitive and spiritual heavy lifting? Descartes thought that the mind was a kind of disembodied space in which representations of things in the world could be manipulated (by disembodied thought processes) and that this process of thinking was the epitome of our higher nature. This mind was, roughly, like a brightly lit executive office in which sat an intelligent little Mini-Me who received information about the state of the outside world from the Perception Robots in the body's sensory systems, made rational deductions and decisions about the best thing to do, and then sent edicts to the Action Robots on how to behave. Everything of importance is available to the scrutiny of the CEO: that's 'Me'. The workings of the mind itself are like the workings of a clock with a glass

22

cover, and the CEO is able to inspect what is going on within. Introspection is possible, and valuable. Descartes wrote to his friend and mentor Mersenne, 'As to the proposition ... that nothing can be in me, that is, in my mind, of which I am not conscious, I have proved it in the *Meditations*, and it follows from the fact that the soul is distinct from the body, and that its essence is to think.'[11]

Thus we have inherited a view of 'mind' as 'the organ of intelligence', and of 'intelligence' as predominantly rational, conscious and dispassionate. Just as the lungs are the organs of respiration, we might say, or the heart the organ of circulation, so 'mind' is used to refer to the whatever-it-is behind the scenes that makes us intelligent. When we do something stupid, we presume that this organ has not been fully or properly engaged, so we are 'absent-minded'. Someone who habitually thinks clearly, or who makes good judgements, has a 'good mind'. In this sense, the word 'mind' often does the job of an explanatory fiction. If we are honest, we don't really know why or how our 'minds' work well one minute and sloppily the next. All we know is that the thoughts we had were either well formed and relevant or random and loose-weave; or that a train of thought preceded an intervention that did or did not turn out to be appropriate and effective. A cardiologist can give a pretty good account of how the 'organ of circulation' works, but after more than a hundred years of scientific psychology we still struggle to give an overall account of this mysterious organ of intelligence.

This picture of the human mind and its relation to the body has influenced and to a large extent controlled society's view of human nature for nearly 400 years. As George Lakoff and Mark Johnson say, 'Descartes has left us with a theory of mind and thought so influential that its main tenets are still widely held and have barely begun to be re-evaluated. It has been handed down from generation to generation as if it were a collection of self-evident truths. Much of it is still taught with reverence.'[12] Still today, clever people can look on their bodies merely as an irritatingly fallible form of transport – a way of getting their minds to a meeting – the maintenance of which takes up valuable time and distracts them from the much more important business of having opinions, winning arguments and making decisions.[13]

Given this image of the mind, it is no wonder that our cultural conception of intelligence focuses on rational verbal and mathematical problem-solving. The typical IQ test takes people out of the context of their normal lives, full of experience, feelings and concerns, and sits them in a strange room that is supposed to be 'neutral' but is, for most people, alien and stressful. There they are asked to solve, against the clock, abstract puzzles that require them to discern and manipulate relationships between out-of-context words, sums and meaningless geometric symbols (see the example below).

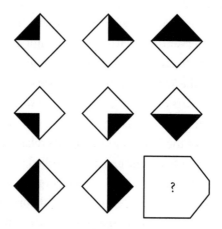

Fig. 1 An example of a Raven's Progressive Matrices type of Intelligence Test.

The mind's I

So deeply embedded is this image of the human mind that it has shaped our sense of ourselves: our identities. When we come to replace the disembodied notion of mind with an embodied alternative – the idea that the mind, the organ of intelligence, *is* the body – we are not just changing theories, therefore, we are challenging and upending very basic, largely hidden assumptions about our very nature. Before we get to the upending, however, I think it would help to have a summary of what exactly it is that is being challenged. What is the Cartesian/Platonic image of human nature? Luckily, cognitive anthropologists such as the University of Connecticut's Roy D'Andrade have done a lot of the work already to surface these assumptions. Here is a rough summary of the main points of the preceding paragraphs that embeds them in a wider framework.[14]

There is a 'Little I' inside me who is in charge – or who tries to be, with variable success. Little I gets a picture on a screen of what is going on (both in the outside world and inside the body). The pictures on the 'screen in the head' are, at best, accurate representations of What's Going On Out There. Usually, the world I see is the Real World. Sometimes I makes mistakes (I thinks you are angry when you are just tired); but mostly I sees things as they are. Little I then blends this picture with its ideas, memories, current job-sheet and arguments. It weighs things up, interprets them, embellishes them, (usually) decides what to do for the best, and then tells the muscles what needs doing and the mouth what needs saying. Little I is not made of Body, though it has a close relationship with it. When the body dies, Little I either dies with it or hops into another body (reincarnation) or decamps to a different realm (e.g. heaven or hell).

(Which one of these you subscribe to depends on which you find most comforting or familiar.)

Little I is the chief Instigator of intelligent speech and action. It is largely responsible for the choices I make. The mind is like a well-lit conscious workspace where Little I does its thinking. Little I can see what is going on in the mind. 'Introspection' is possible and largely accurate. All intelligent thought is conscious and, preferably, logical. Little I is at its best when it is at its coolest. Rational deliberation is the highest form of intelligence. Judges, professors of philosophy and mathematicians epitomise this. We should all aspire to be more like them, consciously weighing up pros and cons and coming to justifiable conclusions before we act.

What goes on in the Mind is – or ought to be – accompanied by a feeling of being in control. If conscious events feel out of control, that is disquieting, and efforts must be made to re-establish control. Sometimes, inexplicably, these efforts seem to make matters worse. (Whatever you do, don't think of an elephant.) To 'explain' waywardness, you can allow a mysterious kind of nether region of the mind called The Unconscious, which isn't Body either but suffers from bodily cravings and beastliness. Sometimes, puzzlingly, The Unconscious bubbles up with quite useful or creative things. In general, waywardness is a glitch in control for which, if I think carefully, I can usually find a simple cause. If I was ratty, it must have been because I didn't sleep well last night. Or because I had had a bit too much to drink.

Emotions are of the body (or the unconscious) and, unless carefully controlled, often threaten to disrupt Rational Intelligence. Emotions are generally selfish and 'primitive'. It was once widely believed that women, being more subject to the tides of emotion, were in consequence less intelligent than men. Little I possesses a fallible instrument for controlling emotion called Will. This is a kind of force that varies between people: some are iron willed and some are weak willed. Some people believe that God can top up your will for you, and being good increases the likelihood that He (*sic*) will do so. People who are weak willed are at risk of being possessed by the Devil (for instance) and being made to do bad things.

Under normal circumstances, minds are individual, self-governing entities. Little I is master in its own house. If I am being influenced by other people I am usually aware of it, and can usually control how influenced it suits me to be. I am independent-minded – unless something spooky is going on like hypnosis or demonic possession. In general I know myself pretty well. I have a reasonably stable set of traits, characteristics, preferences and concerns that constitute my personality. Little I is also the main actor in the continuing drama of my life. The narrative is patchy but largely accurate.

Not all of these beliefs are entirely false, but, as a constellation, says the science of embodiment, they grievously misrepresent human nature and cause us a lot of trouble (which it is hard work to recognise and put right). The rest of the book will try to convince you of this.

Disembodied psychology

The Cartesian view of the mind, with its attendant denigration and neglect of the body, was mirrored – in fact strengthened – in academic psychology throughout the twentieth century. 'Cognition' was the central topic of interest: indeed, from 1955 until very recently, 'cognitive' psychology was about the only kind there was. Sandwiched between Perception on the input side and Action on the output side, Cognition was unpacked, for generations of psych students, into such topics as 'memory', 'comprehension', 'decision-making' and 'problem-solving', all of which were construed in the most conscious and rational terms. Memory was memory for facts – often simply words or syllables which had been randomly paired. Comprehension involved stitching word-meanings together according to syntactic rules. Decision-making and problem-solving pitted students' wits against abstract, stylised puzzles – often versions of the kinds of logical problem that composed most 'intelligence tests' – and compared their actual performance, invariably unfavourably, with a hyper-rational ideal.

There is a widely quoted psychological experiment in which Alexander Luria, a famous Russian psychologist, said to some Siberian peasants: 'All the bears in Omsk [a town a thousand miles away] are white. Someone in Omsk walks out of their house one morning and sees a bear. What colour is it?' Most of his respondents wisely answered that they were unable to say, as they had never actually been to Omsk. Luria protested that he had just told them that all the bears there were white. But we don't know you, the peasants replied. How can we judge if you are telling us the truth, or if you may be mistaken? In the real world, that seems an eminently intelligent response. But not in Logic Land.

The natural allies of cognitive psychologists in those days were logical and linguistic philosophers such as Jerry Fodor, theoretical linguists like Noam Chomsky, who were interested in syntax rather than meaning, and researchers in 'artificial intelligence' like Allen Newell, who were trying to get robots to solve logical puzzles. Neuroscientists and other kinds of physiologists, who were interested in the way real bodies worked, were seen as very junior partners. Once Fodor, Chomsky and their friends had worked out the *design specification* of the mind, then the neuro-people could go off and see how it had been implemented in the 'wetware' of the brain, but that was a rather menial job – merely a kind of confirmatory house-keeping. The fact that human beings were creatures of need, passion and urgent, in-the-moment action, as well as cognition, was not held to be significant.

When the computer became the dominant metaphor for the mind in the 1960s, the focus on logic and language was intensified. Computers 'thought' in terms of logical operations on strings of 1s and 0s – that was their 'machine code' – so the mind was supposed to function similarly. *The Language of Thought*, as Fodor entitled one of his books, used a set of logical or grammatical rules to convert strings of symbols into other symbols – regardless of what those symbols actually meant, or whether they meant anything at all to a real person.[15] These rules were supposed to be innate and universal, and it was the job of 'cognitive scientists' to figure out what they were. On this view, provided the human mind machine could carry out the requisite computations, it was irrelevant whether it was made out of meat, silicon chips or Lego.

Computers offered a tempting metaphor for the divorcing of Body from Mind in the form of the distinction between hardware and software. We were encouraged to see our brains as composed of multipurpose biological hardware on which could be run a host of socially transmitted software programs and belief systems. But this is quite misleading. Computers have no genes and no inbuilt needs and concerns. They are truly cognitive devices, devoid of inherent feelings and motives. Remember also that the computer model of mind was shaped in the days before networking and the internet, so minds, like those early machines, were seen as individualistic and self-sufficient. They were fed problems, they computed on their own, and they produced answers.

So human beings, cast in the mould of the early computer, and pared down to their most rational and abstract capabilities, were deprived of almost everything that made them rich and interesting and real. Anything to do with psychoanalysis, or feelings, or friendships, or even ordinary interests like football or fashion, was treated with a neglect that bordered on disdain. In my days as a young graduate student at Oxford in the early 1970s, I became involved in interviewing sixth-formers who wanted to read psychology. The only bit of advice I was given, for this important task, was to ask them if they were interested in 'what makes people tick', and if they said they were, to reject them.

Perhaps this sense of the brain and the mind as being fundamentally rational reflected the mindset of philosophers and scientists themselves. After all, the people who thought and wrote most about the brain-mind were people whose stock-in-trade was formulating clear propositions, constructing watertight arguments, drawing verifiable inferences, and reaching logically sound conclusions, so it was no surprise that the kinds of thinking they were particularly good at turned out, in their theories, to be the kinds of things that brains were fundamentally designed to do! Even before the rational age, writers of non-fiction tended to be those whose bent was to argue and reason, so we might well expect them to have left a gimpy set of tracks in the snow of history.

However, there is no reason why such 'high rationality' should constitute the fundamental *modus operandi* of the brain. It is perfectly possible that brains naturally function in a profoundly *un*-philosophical way, and yet, *if well trained*, have the capability of serving up products that are philosophical. Just as you can now program a digital computer to behave as if it were a brain, so cultures have for two millennia been programming human brains to simulate the rationality of a machine. As a case in point, Peter Medawar and Paul Feyerabend showed many years ago that the actual way in which scientists work is completely different from the neat, logical picture that the published scientific paper suggests.[16]

*****

Overall, the body has had a hard time of it for the last 2,500 years of human – especially 'Western' – history. Because the world has lacked, until very recently, the scientific tools to persuade the body to reveal its intricacy and sophistication, it is no wonder that societies should have invented all kinds of explanatory fictions to account for human intelligence, leaving only menial tasks for the body to pick up. No wonder the body became a Cinderella concept, denigrated and disdained by the fictional Ugly Sisters of Mind and Soul. For 2,500 years people have been inveigled into believing in a ghostly world of Platonic abstractions and theological idealisations, and encouraged to aspire to a dispassionate way of thinking. It's not just that dispassionate thought is sometimes a useful adjunct to bodily feelings and intuitions; no, Reason is Better, period.

But now we do have those tools, so let's put them to work. Let's take a more detailed look at the human body and see what it is really like.

3

BODIES

WHAT ARE WE MADE OF?

I am body entirely, and nothing beside.

Friedrich Nietzsche

Let's begin by taking a quick trip around the human body. I'll often call it 'your body', or sometimes 'my body', because this is not a matter of remote academic interest; I'm talking about what's inside you and me, and how it is working, right now. I'm a scientist and, at least in this regard, a Nietzsche man, so I start by assuming that everything we do and think emanates, somehow, from this sophisticated biological construction. The eyes that are reading these words, the brain that is having thoughts in response, the facial muscles that are twitching involuntarily to signal agreement, amusement, puzzlement or irritation, the way those reactions are affecting the digestion of your last meal . . . all of this and much more adds up to who you are, right now.

So let's have a look at what stuff we are made of, and how it is behaving. Unfortunately I can't just ask you to look directly at what is going on, because we don't have direct access. Much of what is happening inside us is not accessible to consciousness – you can't feel the red blood cells being manufactured in your bone marrow as you are reading – so we will need the help of science to get a handle on what bodies are and how they work. I'll briefly introduce a number of complementary perspectives.

You are a bunch of cells

What are little boys made of?
Slugs and snails, and puppy-dogs' tails.
That's what little boys are made of.
And what are little girls made of?
Sugar and spice, and all things nice –
That's what little girls are made of.

Nineteenth-century nursery rhyme

Your body is the result of an evolutionary decision by living cells, as Benjamin Franklin put it, to hang together rather than to hang separately. The story of the evolution of multicellular organisms is the stuff of school biology these days, so a quick reminder will suffice. Somehow, around four billion years ago, some molecules came into being that had a curious property: they were able to reproduce themselves. More than that, they seemed bent on doing so. After a while, these replicator molecules had become able to create tiny homes for themselves that suited them and their reproductive tendencies. They wove membranes that enabled them to construct their own miniature ecosystems and so were able to keep conditions more or less to their liking. Millions of descendants of these first single cells – bacteria and microorganisms of many kinds – are swarming inside you and me right now. They live symbiotically with us, often to our benefit, and sometimes, when they make us ill, to our detriment.

Some of these cells set up working arrangements with even smaller cells. Mitochondria, for example, are simple unicellular life forms that have the deeply beneficial ability to generate energy, so our human cells welcome them as long-term lodgers, feeding and protecting them in return for a constant supply of fuel. Some of the cells learned how to divide, so they were able to replicate not just their DNA but the favourable cellular living conditions as well. And some of these tiny creatures discovered (by natural selection, over many generations) that, having split, there were advantages in sticking together, especially when they learned the art of specialisation, so that different cells could do different jobs on behalf of the whole community.

The survival strategy of forming collectives has both pros and cons, so not all multicellular experiments survived and not all cells decided to go down the path of living as a commune. An amoeba is vulnerable, but its needs are simple. If its world stays stable, it can easily get by on its own. It is low-tech and low-maintenance. But if the world changes, and what it needs stops floating by, there is little it can do. Big bodies, on the other hand, can move further and faster in search of food, can spot threats coming earlier, and can defend themselves more skilfully and vigorously. But they are expensive to maintain, and there is a great deal more to go wrong. Feeding becomes a military operation, requiring split-second timing of fork-raising by hands, mouth-opening by muscles, chomping by teeth and lubricating by saliva, swallowing by more muscles (making sure you also stop breathing at the crucial moment), squirting by digestive chemicals and churning by the stomach, and so on (you can imagine the rest for yourself). And communal living is not a breeze, as many of us discovered when we were students. Getting the right balance between 'What is good for Me' and 'What is good for Us' is often tricky.

34

Much of the detail of this evolutionary story is not essential here, but one thing is. The cells of which we are composed, individually as well as collectively, act as if they want to survive. This doesn't mean they have conscious desires or preferences; it means they have a set of built-in responses to potentially adverse events that tend to neutralise or avoid the damaging consequences of those events. If the internal milieu of the cell goes off kilter, that by itself triggers processes that try to restore the balance. If their skin is punctured, things happen that have the effect of repairing the puncture. Cells that have these self-protective reflexes survive longer, and therefore have better chances of replicating themselves, than those that don't. At our most basic level, we are purposeful – and intelligent. As Antonio Damasio says in his book *Self Comes to Mind*, 'embodied knowledge of life management precedes the conscious experience of any such knowledge. [This] knowledge is quite sophisticated . . . its complexity is huge and its seeming intelligence remarkable'.[1] (I just wonder why he felt the need to pull his punches by inserting that 'seeming'.) I think this built-in disposition and capability to act in a way that meets your needs and concerns – especially as those concerns proliferate and your world gets more complex – is the very heart of intelligence.

You are a form of motion

Next to her the warm body [of her dog] shifted slightly, and she wondered if she was giving off minuscule tensions that disturbed sleep. She was trying to remain motionless, but that, as a kinesics expert, she knew was impossible. Asleep or awake, if our brain functioned, our bodies moved.

Jeffrey Deaver[2]

A while ago, I was sitting in the grand foyer of Sharpham House in Devon in the presence of a hundred or so other living beings and the dead body of Maurice Ash whom we had all known, admired and loved. I was watching his body intently, and I saw him take a breath. I actually saw his chest move. Of course he didn't really breathe, but my brain so rebelled at the utter stillness of the body that it overrode the facts and, for a moment, insisted on bringing him back to life. Complete stillness is incompatible with life: it is anathema. A child that is born still makes one revolt at the wrongness of it.

The body isn't a thing, it's an event. We exist by happening. When we look in the mirror, we see a familiar object looking back. When we look down, our legs look and feel solid enough. But we know that if we stop happening, we quite quickly start to fall apart. If we lock up the house and go on holiday we imagine it will just stay put, but not so the body. Even when we are asleep neurons are firing, cells are developing, bladders are filling, blood is pumping, lungs are breathing, legs are twitching and throats grunting. In a thousand ways, large and small, we are squirming and wriggling the whole time. Moving is not something we do from time to time and rest in between. Rest is itself a form of motion. There is special circuitry in the brain, called the default network, which starts up precisely when we have nothing urgent to do. On the micro level, we are constantly abuzz with activity.

If we don't do things, we don't survive. Human bodies, like their individual component cells, are constantly replenishing their nutrients. In order to provide our cells with the necessary raw ingredients, from time to time we have to get up and go to the fridge or phone for a pizza. To be alive is to need supplies. And to get them, we need to be mobile. So we have evolved to be active at the macro, behavioural level, as well as at the micro. In the long run we will lose the battle for survival, but until death our complex, multicellular bodies, like the single cells from which they are made, act as if they want to keep on living – till we have done our reproductive duty and more. Our bodies' chief priority is action in the service of their survival, well-being and reproduction. As neuroscientist Daniel Wolpert puts it:[3]

> While sensory, memory and cognitive processes are all important, they are only so because they either drive or suppress future movements. There can be no evolutionary advantage to laying down childhood memories or perceiving the colour of a rose if it doesn't affect the way you are going to move later in life.

In short, we are restless, active, proactive creatures, full of hopes, desires, fears and expectations; bundles of projects, short term and long term, worthy and unworthy, humdrum and grand. As I sit here, I am interested in what the boys on the beach are doing, shifting my posture to get comfortable, thinking about how to cook my chop for supper, wanting to finish the chapter, searching for the right word, remembering I need to water the plants on the balcony ... and on and on. Biological intelligence can't be understood unless it is built on this core concern with matters of action and desire.

As I said in the last chapter, we aren't like computers: machines happy to wait patiently until they are switched on and told what to do. We are constituted by a fluctuating (and expanding) portfolio of inherent concerns, and constantly animated by physical attempts to address those concerns. We might note that the word 'mind' itself still carries, in everyday speech, strong echoes of that motivational impetus. To *mind* is, at root, to care. If you *mind* about the result of the match, you don't just think about it, you are concerned about it. It matters.

You *mind* terribly – as opposed to your partner who may not *mind* at all. A child-*minder* is not just conscious of the toddler; that awareness is imbued with concern. We *mind*, in the sense of care, before we are minds, in the sense of possessing organs of intelligence. Fundamentally we are not designed for thinking, philosophising or solving cute logic problems against the clock. Reason and debate are themselves tools that evolved to support deeper biological agendas.

The reason we have a brain is mainly to figure out and implement the right movement to make, given our current concerns, and this is a complicated problem. Animals that move have brains; those that don't move, don't. There are a few animals, like the sea squirt, that start out as movers – they swim around – and so possess a basic brain for coordinating their movements. After a while, however, they decide it would be less effort to spend the rest of their lives as plants, so they find a good spot, put down roots, and survive on whatever comes their way. Having no further use for their brain, they eat it. (As the old academic joke has it, this is not unlike what happens when university professors finally get tenure.)

You are a flexible cage

The physical manifestation of the body is primary. The stuff of intelligence has evolved in conjunction with that body, and is a modulator of its behaviour, rather than a primary and central control system.

Rodney Brooks[4]

The first intelligent thing that mobile creatures have to have is a body that moves in the ways they need it to. Some just have little tails they can swish around, and others are able to shrink away when prodded. But the lifestyle of our multicellular forebears needed more speed, more strength and more precision than that, so they evolved a skeleton. This is a semi-rigid but partly elastic cage made of bones that protects and supports the soft organs within. It has joints to which limbs are attached. Bones are connected to each other by strong bands of elastic muscles which can contract, causing the bones to move relative to each other. Muscles are attached to bones by tendons and bones are connected directly to one another by ligaments. Evolution has adjusted the lengths, weights, arcs of movement and elasticity of this articulated cage so that the skeleton as a whole can move quite smartly, even without much input from a brain.

When you were young, you might have had (as I did) a simple toy that would walk by itself down a slope. If you got the slope at the correct angle, it would walk steadily down without either stopping or falling over. It did so just because of gravity and the mechanical way it was constructed. It would start to topple forward, but that toppling freed one of the legs to swing forward like a pendulum. Still toppling, it would stand on the first leg and thus free the other to swing forwards, and so on. Now, if you want to make a version of that toy that would walk on the flat, you have two options. One is to start from scratch and design a coordinated set of sensors and motors to move the legs in turn. For that you would need a small brain. Alternatively, you could design a way of getting the toy to tend to topple again – perhaps by giving it a long and heavy nose – and then rely on the same cheap, mechanical pendulum effect of the legs to keep it walking. You might have to adjust the parameters a bit, but it would work.[5]

Rolf Pfeifer and his colleagues at the Artificial Intelligence Lab at the University of Zurich have explored the potential of the second strategy in some detail.[6] Instead of loading the intelligence into a kind of brain that controls the leg movements 'from the centre', they have shown how much of this intelligence can be embodied in the physical make-up of the body – if you get all the weights, resistances and 'springs' (muscles) right. The figure below shows Puppy, one of their four-legged robots built on these principles. Puppy walks on the flat in a lifelike way, but it has no brain: no organ of central coordination.

To get more sophisticated behaviour like running uphill, or walking on uneven ground, you do have to introduce some brain-like control, but only to moderate the behaviour of the 'intelligent skeleton', or to kick it off: not to orchestrate everything. A robotic relative of Puppy, called BigDog, is able to walk on slippery or muddy surfaces, for example, and can pick its way over a heap of rubble without falling over. (You can see BigDog in action at www.youtube.com/watch?v=cNZPRsrwumQ. It might better be called BigFly as the go-cart motor that drives it along emits a very loud un-canine buzzing noise.)[7]

These K-9s illustrate an important characteristic of how real bodies work; a good deal of the apparent intelligence of the

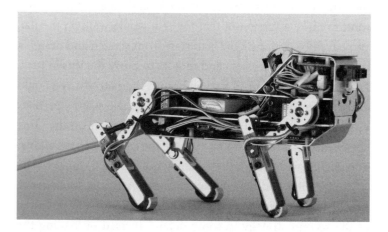

Fig. 2 The robotic Puppy.

mind and the brain is distributed around the physical systems of which the body consists. Getting Puppy and BigDog to walk may seem somewhat trivial compared with the intricacies of mind and body for which the accolade 'intelligent' is usually reserved, but we will see how similar principles of embodiment continue to apply as we scale up. Pfeifer sums up this important insight thus:

> Behaviour is the result of an agent interacting with the real world, which includes not only the agent's neural system but its entire body: how the sensors are distributed, the material properties of the muscle-tendon system and the joints, and so on. This collection of interdependent mechanisms [we can call] the agent's embodiment.[8]

The idea that intelligence can be embodied in physical structures, and that such structures can therefore take some of the strain off minds and brains, is a key one in the science of embodied cognition.

*****

If we are trying to creep up on the idea of 'intelligence', we cannot leave the subject of the musculoskeletal system without saying a word about the human hand. Without the hand we would not be able to make or use even the simplest tool, and without tools, we would demonstrably be a whole lot less smart than we are. As Raymond Tallis puts it, in typically droll fashion: 'If Adam and Eve had been expelled from Paradise with paws instead of hands, the history of the human race would have been unimaginably different.'[9] Where would we have been without our ability to grasp, caress, pluck, catch, pull, twist, pinch, prod, punch, rub, scratch, tap, drum, throw, write, squeeze, tickle and dozens of other manually clever things?

Undoubtedly the human hand requires a sizeable brain to accompany it. Indeed, many, like Tallis, have argued that the evolution of the human hand (as opposed to the paw) may well have been one of the principal drivers for the evolution of the brain. But, as with the example of walking, we must not assume that the hand is just a highly articulated but essentially passive tool that depends on an intelligent brain to tell it what to do. Some of the 'joint intelligence' that the brain+hand so obviously possesses is embodied in the hand's physical composition.

The hand has a thumb that can swivel, and which is also of just the right length to enable it to 'oppose' the first finger. This enables us to pinch and grasp. At the base of the thumb are unusually strong muscles, so that the precision of the grip can be backed up with considerable power. We have flexible, jointed fingers that can (and do) curl up, and also move independently of each other (to some degree). We have nails instead of claws, and these provide a protective backing to fingertips of extraordinary sensitivity, capable of detecting tiny changes in texture and resistance. We have ridged skin on our hands that increases friction, providing increased grip for larger objects. The skin is also capable of being deformed and compressed like a cushion. And the whole constellation of small bones, muscles and tendons means that when the hand starts to close around an object its physical construction will automatically adapt its grip to match the shape of the object – without any help from eyes or brain.

Japanese robotics researcher Horoshi Yokoi has created a prosthetic hand with many of the same physical properties as a real one, and has shown how it can grasp a wide range of objects very successfully without complicated control. You just tell the Yokoi hand to 'close', and physics does the rest (see Figure 3). This

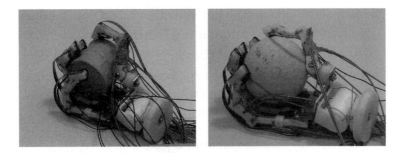

Fig. 3 The Yokoi hand.

is good news for people who have prosthetic hands. Their brain can produce motor commands that can be picked up on the surface of their skin by an electromyograph (EMG) and converted into control signals for a robotic hand. But EMG signals are very 'noisy' and therefore control of a conventional prosthesis, which requires precise signals to tell it what to do, is poor. The Yokoi hand performs much better – and is a lot cheaper.[10]

When we come back to the human body as a whole, we find another set of problems to do with general coordination. The musculoskeletal system has many movable parts, and they must work in concert. Some of this needs delicate central orchestration by the brain, but again the brain modulates what the body is doing rather than having to design it from scratch. One of the ways it does this is through the *physiological tremor*. As far back as the 1880s, it was known that the entire muscular system of the body was constantly vibrating at a rate of around 10 cycles per second. When any part of the body moves, its activity is overlaid on this tremor, and this helps the rest of the body to stay coordinated – just as a pair of dancers can be better 'coupled' when both are listening to the same music and their movements are tied together by a common beat.[11] This example reinforces two of our main points in this chapter: the intrinsic activity of the body, and the way in which this inherent activity contributes to the overall intelligence of the whole person. The brain cannot be properly understood except as one element of a larger system that includes the body.

You are your organs

Inside and around the skeletal cage, and beneath the protection of the skull, are clustered the major organs, substances and systems of the body. Each of these carries out functions that are necessary to keep us alive. Life, even as a multicellular being, is precarious, made possible only when a large number of conditions are met simultaneously within the body.[12] The body can only tolerate small fluctuations in a whole range of its parameters. For example, the composition of various gases inside us, the acidity of the inner fluids that constantly bathe our cells, and the temperatures that are conducive to the chemical reactions necessary for survival can vary only within narrow limits. The major organs of the body are each designed to keep track of a subset of these variables, and have ways to bring them back to optimal values if they begin to wander off. When energy levels flag, we have to locate food, get it inside us, convert it into the universal currency of energy, the ATP molecules, get the energy molecules distributed to wherever they are needed, and get rid of the waste products. The internal division of labour means that clusters of cells devote themselves to these different but interlocking tasks, thus constituting various 'centres of operation', which we carry around inside us.

If we are to uncover the intelligence of the body, we will have to overcome our Platonic squeamishness and dive into our moist innards. A venerable Buddhist meditation may help here.

It is called *Patikulamanasikara*, which roughly translates as 'reflections on repulsiveness' – and it is that attitude of anxiety about and disgust at our own substance we need to challenge. Just because our insides are dark, slimy and sometimes smelly, that doesn't make them 'dirty' or 'bad'. *Disgust* at our own insides only arises within the historical framework we discussed in the previous chapter. The meditation lists the physical ingredients of the body. They are (in slightly updated form):

Heart, lungs, spleen, lymph nodes, pancreas, kidneys and liver
Brain, eyes, ears, nose and tongue
Skin, flesh, bone and bone marrow
Stomach and the rest of the digestive system
Hair (in various locations), nails and teeth
Skin, tendons, ligaments, cartilage, gristle, diaphragm and nerves
Faeces, urine, sweat, grease, fat, semen, snot, phlegm, saliva, tears, blood, lymph and pus

Monks are encouraged to (as one handbook puts it) 'contemplate the body ... as being full of many impurities' – but the point is not to cultivate that reflex of disgust but to learn to replace it with an attitude of polite interest. Innards R Us; they are our family; they need only make us anxious when they become visible. *Concern* when those insides spill out as vomit or blood is evolutionarily entirely appropriate. But if we are to really understand the body, and learn to be it, we have to befriend it, locks, snot and marrow.

When we think of our insides it is often the stable, visible structures we focus on – brain, heart, lungs, kidneys, liver, intestines and so on. Because they can be readily dissected out of the body, they have tended to be treated in terms of their most evident functions, and understood through simple analogies. The lungs are like bellows; the heart is like a pump; the stomach is a food processor; the kidneys are filtration plants for the body's fluids; the liver is a pharmaceuticals factory. And indeed they do perform these functions, and much besides. Each is 'intelligent' in its own right. But when we come to think about the intelligence of the body as a whole, it is the way the different organs and elements communicate with each other – how the intelligent whole becomes more than the sum of its parts – that is of more interest.

Take the heart, for example. In 1991, Harvard physiologist Ary Goldberger reported an intriguing finding. As we have seen, the different bits of the body are generally thought to have an ideal level of functioning, and if they are perturbed from this – if blood sugar falls below this level, or if blood pressure rises – then they start to behave in a way that gets them back to the 'ideal point'. But how precisely defined is this optimal level? Is it a point, or is it a range? It has usually been assumed that it is the former; if everything is fine, there is no need for the heart rate, for example, to vary at all. The healthy, happy heart should have a constant beat. But Goldberger found that it doesn't. ECGs of a normal heart show that the rate varies quite a lot when there seems no reason for this, and it does not vary in any obviously regular way. Hormone levels in the blood serum of normal, healthy people also fluctuate much more than seems necessary. So what is going on?[13]

The answer seems to lie in the fact that the heart, if it is to stay healthy, needs to keep talking to the rest of the body. It is not just doing its thing in isolation. The heart actually has its own little brain, which enables it to keep in constant communication with the lungs, the liver, and the rest of the central nervous system. There are inputs from the sympathetic nervous system that normally increase blood pressure and/or speed the heart rate up, and other inputs from the parasympathetic nervous system that generally lower Blood pressure (BP) and/ or slow the heart down. And there are outputs from the heart that send messages about its current mechanical, electronic and chemical state to other organs and upwards through the autonomic nervous system to the brain stem, and thence to the brain as a whole. In between these incoming and outgoing neurons, there is the heart's own web of internal communication neurons that relay the news that is coming in and modify the information it is broadcasting. The heart is an inveterate tweeter.

This 'little brain on the heart', as Canadian researcher Andrew Armour calls it, uses all this activity to modulate the heart rate, and other parameters, on a beat-by-beat basis.[14] In other words, the healthy heart beats somewhat erratically because it is in constant resonance with the wider body of which it is a part. Homeostasis prescribes not an ideal 'sweet spot' that the heart is forever trying to return to, but a 'sweet zone', a range of values, within which it must have the freedom to move around. And indeed, one of the characteristics of the unhealthy heart is a loss of this elasticity, and an increasingly regular heart rate. Goldberger has found that sudden cardiac arrest – a fatal heart attack in an apparently healthy heart – is very often preceded by an increase in the regularity of the heartbeat. Lacking its usual inherent flexibility, the heart cannot respond as effectively when the conditions around it suddenly change. It is as if it has stopped listening to the radio, so news of an impending tsunami goes unheeded and the heart is therefore caught unawares.

Other bodily systems also have brains of their own. The digestive system contains its own nervous system that enables it to regulate its responses to inputs of various kinds, even when it has been disconnected from the brain and the spinal cord. Evolutionarily this is not surprising. We were digestive tubes long before we developed separate brains – as each one of us was, once upon a time, in the womb. The body has been well described as a dense city built along the banks of a busy waterway, the alimentary Grand Canal. Goods arrive at landing stages, fish are caught along the banks of the canal, and waste is discharged downstream. Regulating the transactions along the margins of this tube is a complex business: there are more than a billion nerve cells in the small intestine alone. Many are concerned with the activity of the gut itself, for example in maintaining healthy relations with the vast army of bacteria that have set up camp principally in the lower gut. But many are also in constant communication with the other organs and regions of the body, including the brain.

You are biochemical soup

As well as the solid organs, you are composed of other systems that cannot do their job unless they are floating around, permeating the body as a whole. The circulatory system is one, carrying the energy-giving red corpuscles of the blood to the remotest corners of the empire. But the fluid of the blood, the plasma, along with the lymph system, also carries a host of chemicals that help with self-defence and self-repair. The liquid immune system carries the so-called white blood cells, technically the lymphocytes, which are a kind of 'homeland security' service, constantly checking the papers of the molecules and microbes they meet to enable the body to tell friend from foe. Each lymphocyte has a big sticky molecule, an antibody, attached to its surface, which is like a Wanted or Missing Person poster of the particular friend or foe for which it is on the lookout. When it finds a match, it either gives it a hug or performs a citizen's arrest. The lymph nodes dotted around the body are local jails and interrogation centres for any undesirables that get picked up.

The traditional image of the immune system is as 'civil defence', but it is much more than that, because the friendly interactions turn out to be at least as important as the hostile ones. The lymphocytes are not just the secret police; they are a vital mobile throng of meeters and greeters, exchanging information and keeping the different regions of the body literally 'in touch' with each other and with the whole. The Chilean immunologist Francisco Varela, whose work has been seminal in the development of the whole area of embodied cognition, says that the immune system is really there to keep knitting the various specialist groups within the body into a single somatic identity.[15] When the immune system spots intruders, it has a sophisticated capacity to deal with them. But this is only how it responds to emergencies. Most of the time it is monitoring and maintaining the body's sense of coherence and identity, and this is key to how we function in the world. Mice that are raised in completely sterile environments still develop a nearly normal immune system, for example, because, even without the threat of disease, the immune system is essential. It thus has to keep up a constant dialogue with the brain. As Varela says: 'A more sophisticated psychosomatic view will not develop unless and until we understand that the immune system is a cognitive device in itself.' It is part of our intelligence.

Intermingled with the lymphocytes and erythrocytes (red blood cells) floating around the canals and marshes of the body's interior are a host of chemicals that help to regulate what is going on: the ingredients of the endocrine or hormonal system. They include neurotransmitters and neuromodulators – chemicals that regulate the firing of neurons – such as adrenaline, acetylcholine, dopamine and serotonin; sex hormones such as oestrogen and testosterone; and a wide variety of short amino acid chains called peptides, for example insulin which regulates the metabolism of carbohydrates and fats; cortisol that affects the way the immune system responds to stresses of various kinds; and oxytocin which alters mood, especially in relation to caring and intimacy. These hormones are manufactured in various glands and released into the bloodstream, into the lymphatic system or directly into the free-flowing fluid that surrounds all the cells. All of these floating chemicals carry news, advice and instructions. Hormones, for example, have a specialised molecular 'smart card' that will only 'work' when it finds the corresponding 'hole-in-the-wall' slot embedded in the wall of one of the cells which it happens to be floating past. When a fit occurs, the cell is triggered to change its behaviour in a particular way.

So these organs and systems are not functionally separate from each other. When I was a student, I was taught that the body had three distinct systems, the endocrine system that circulated regulatory hormones, the immune system that fought infection, and the nervous system that sent electrical messages around (of which more in the next chapter). But not only has our understanding of these systems developed enormously over the last thirty years; we now know that it is much more accurate to see them not as independent and parallel, but as aspects of a single system that looks after many complementary aspects of our well-being.[16]

The nervous system is as much a chemical system as an electrical one. Neural impulses are triggered by microscopic changes in chemical concentrations, and regulated by all kinds of chemical neuromodulators, which may originate in a wide variety of organs and tissues. The constant neurochemical tweeting of the heart is just one example of the constant cross-talk between different bits of the body. The lungs and the heart are talking to each other via the hormones and antibodies coursing through the pulmonary artery. The immune system penetrates into the innermost recesses of the digestive system, and interacts with the neurotransmitters in the gut's own nervous system.[17] And so on. In addition, we might add that all of this activity is affected by the dynamics of the muscles and joints. Large-scale physical activity – from the runner's 'endorphin high' to the soothing release of oxytocin in the suckling baby – serves to regulate and coordinate what is going on at more microscopic levels.

You are a CADS

What this all adds up to is radical shift in the way we look at our bodies. For centuries we have taken the vocabulary of medicine and used it to think about how our bodies are structured. But it turns out that the language of distinct organs and systems does not do justice to the coherence and interwoven nature of the body. More than that, it blinds us to the essentially systemic nature of our bodily selves. Western medicine has made huge progress by conceiving of the body as an assemblage of parts, each of which can be treated separately when 'it' malfunctions. But we are not built like clockwork, and conventional, so-called allopathic medicine is only just beginning to recognise maladies that cannot be localised to a single source or cause.

Of course there are many other medical systems, such as homeopathy, acupuncture, and those of traditional Chinese herbalism or Indian Ayurveda, that do recognise the intricate interdependencies of the body. But they are often such a tangle of accreted folklore (some of which bears scientific scrutiny and much of which doesn't) that it is impossible to sort the demonstrable truth from the passionately held and anecdotally supported belief. (And, of course, the placebo effect is still worth having.) Out of the kind of work that I am describing here, a scientific framework for understanding holistic effects may well emerge, but hybrid disciplines like psychoneuroimmunology are still in their infancy.

But this is not a book about medicine. The point I want to emphasise here is the more general one that our bodily selves are *systems* – technically, Complex Adaptive Dynamic Systems, or CADS for short.[18] Systems theory is one of the foundation stones of embodied cognition, so it is worth having a quick look at its main features. In many scholars' hands, systems theory quickly becomes pretty technical and often mathematical, so I am going to avoid many of these intricacies here and just try to give a flavour of the approach. I'll call a system in the technical sense a System to distinguish it from more everyday usages (for example, when I mentioned different medical systems in the previous paragraph). A System is a constellation of processes that are themselves Systems: I'll call these contributory Systems, Sub-Systems. When a bunch of Sub-Systems get together and interact with each other in complicated, reverberatory ways, they can create a larger System that seems to have a kind of stability and integrity. It may look like a reasonably stable structure, for example, though this kind of solidity is maintained only through constant dynamic interactions of the Sub-Systems.

A termite colony or a beehive is a System. Hundreds of small creatures – Sub-Systems – are doing their own thing, but they are also interacting through local messages. These kinds of small-scale interaction can lead to the emergence of coherent 'swarm intelligence'. For example, a termite mound is like a Gaudi cathedral, a complex architecture of pillars and vaults, which looks as if it would need high-level thinking and lots of meetings to accomplish. But that's not how termites do it. Individual termites like to make little mud balls that carry a chemical scent, and to roll them around. They deposit them where the scent is strongest – which leads them to create piles of mud balls that compact together into pillars. If two pillars grow up near each other, a termite carrying its ball up one pillar will get a waft of the scent from the adjacent pillar, and so tend to deposit its ball on the side of the pillar nearest to its neighbour . . . and after a while, the two pillars lean together and form an arch. No termite has a plan. No individual knows about arches or intends to build one. Yet it is hard to resist the idea that there must have been intention or design somewhere. The System-level intelligence emerges naturally from a lot of Sub-Systems talking to each other and following their own local rules.[19]

The body is a System in the same way. Its minute elements respond to each other in a way that makes coherence, and the appearance of 'purpose', emerge. Each of the Sub-Systems behaves as it does partly out of its own nature, so to speak, and partly as a result of the interactions it is taking part in. The 'bits' are in constant resonance with each other, and these resonances modify their apparent nature. The bits may seem to be anatomically distinguishable, but they can't be dissected out and expected to behave in the same way. The heart that marches only to the beat of its own internal drum is a sick heart, remember. In a System, any apparent boundary or membrane that seems to mark the limits of a Sub-System isn't a barrier or a stockade; it is a site of constant, vital interaction. If nations retreat into protectionism, and international trade stops, then life within the national borders must adapt. Under siege, life has to change or it stops.

Now, what is sauce for the Sub-Systems is sauce for the superordinate System – because the System is also a Sub-System within a wider set of forces and interactions. As my heart is to my body, so my body is to the world around me. So each System is only the way it is because it is an aspect of a Super-System. The heart beats as it does because it is listening to the rhythms and cadences of the gut and the lungs. I am as I am because I am constantly being licked into shape by the air I breathe, the food I digest, the birdsong in the garden I can hear and the shifting quality of the relationship I have with my wife. I am different when we are not together, and so is she. Judith and Guy are not identifiable 'players' in this relationship any more – when we are together, there is only Judith-in-the-context-of-Guy and Guy-in-the-context-of-Judith. As I move around, the nature of the Super-System around me changes, so I am always Guy-in-the-context-of-something. The relationship has emergent qualities that are not reducible to separable qualities of us as individuals; and we as Sub-Systems are constrained and shaped by the Super-Systems of which we are currently elements. There is no 'But who am I really?' because if I were truly isolated from all the big Systems around me, I'd be dead (or dying). Everything in the living world is a Midi-System, a transiently stable yet always changing configuration of material in motion.

So from the CADS perspective, the human body is not a noun, it's a verb. We aren't like billiard balls that meet, collide and ricochet off unchanged. We are confections constantly being whipped up by a combination of the Super-Systems in which we are participating and the Sub-Systems of which we are composed. We are like whirlpools and eddies in a river that cannot be taken home in a bucket. We are like clouds and waves, constellations of processes that, for a while, have the appearance of being 'things'. We are like gyroscopes or spinning tops that have stability, actively resisting being knocked off course only because they are constantly being spun. If the process of spinning is not maintained, the apparent 'desire' to resist being perturbed disappears. The coherence of bodily structure and behaviour reflects the constant internal resonance of all their ingredients with each other – and with the wider set of Systems within which they are embedded. Out of all this dynamic reverberation emerges a person making a sandwich, and reading a brochure for a new car while eating. Or a baby yelling in the night. Or a heptathlete hefting her javelin.

As we find out more about how bodies are constructed, and how they really work, we cannot help but be impressed by their intelligent bioengineering. We have seen that, without a brain, our bodies are able to do some pretty sophisticated things. So the obvious next question is: why *do* we have brains?

4

WHY THE BODY NEEDS
A BRAIN

Brain, n. An apparatus with which we think we think.

Ambrose Bierce[1]

As bodies get more complicated they need ways to coordinate what is going in their different limbs and organs. But they also need to coordinate all of that with what is going on around them. In order to behave intelligently, we have to be sensitive to what is going on in the Super-Systems that surround us. To do that, we have evolved a range of ways of resonating with this big wide world: our special senses. And the most important of these is touch.

Skin

Everything occurs on the skin.

Hermann von Helmholtz

Skin is incredibly multi-talented. It is the leather pouch that stops us spilling out. It is also an important part of the immune system, being (usually) our first line of defence when toxic or dangerous things come our way. Its pores allow us some measure of temperature control through, for example, sweating and shivering. And skin protects us from the harmful effects of sunlight by producing the dark pigment melanin, which absorbs and dissipates ultraviolet rays. But skin is also our primary organ of touch. It is the original way in which all animals get information about their surroundings. It is our largest and most important sense organ. Skin takes up around 18 per cent of our total body weight, and it is worth every ounce.

Skin is at once our most basic and most sophisticated organ, and touch is our prototypical sense. We could not live without it, and the other senses are merely specialised forms of touch. One can live a rich life without vision, but without touch one is really in trouble.[2] In its early learning a child relies heavily on touch: it feeds, sucks, clings, gets swung around, and later stands, walks, builds, falls and stands again. It never outgrows learning by moving and touching, even when language, imagination and reason have kicked in. It will learn to throw and catch a ball, ride a bike, dance, get on and off escalators, lift suitcases, feel the forces of a cornering car, make love, and swing its own baby in turn.

Touch is an action. We are 'in touch' with the world by moving against it, and feeling it on our skin. Touch occurs when skin and world move relative to each other, and that gives us useful information. We know 'softness' through the activities of squeezing or stroking, and 'hardness' through the physical impact of a fall or a cricket ball. The kind of touch tells us whether to approach or avoid. A punch, a rasp or a prick, and we draw back; a caress, and we snuggle in. Particularly in sensitive areas such as the hands or face, but in fact all over, the skin is densely provided with neurons that specialise in different kinds of touch. One set of nerve fibres, for instance, responds to pain, pressure and temperature as well as itches and scratches. It tells you that an object with a certain size, temperature, velocity and texture is 'in your face' or 'on your back'. A different set responds selectively to touch that is slow and firm but gentle. Interestingly, these so-called C-afferent neurons[3] take a different route to the brain, joining up with afferent fibres from the visceral core of the body, and thus acquiring a strong emotional tone to do with affection and security. We have a whole network of nerves designed to tell us when we are in touch with another human being who is safe and nurturing.[4]

We know the world, to a considerable extent, by remembering how it will feel if we stroke it, pinch it, prod it or tickle it. Through active touching we get the world to reveal itself, and we register those discoveries. If we were blind, we would depend on exploring the solid world in this way, and thus build up a network of expectations: if I moved my hand here, I would expect to feel the hard edge of the table; if there, the cool smooth curve of a bottle. Even on the brink of sleep my body expects that an arm stretched out thus should encounter the soft shape of your hip so. I am not conscious of this tissue of physical expectations and correlations, but I know they are there because when the hip is missing I become alert, and need to go and see if you, unable to sleep, are safely (if grumpily) drinking tea in the kitchen.

This is how all our senses work: by generating these interwoven webs of expectation that link movement to sensation. Even vision is like this, though it is less obvious. The world in the form of electromagnetic information rubs against our moving eyes, and from this I learn that, if things stay still, a flick of my eyes over there will result in a predicted shift in visual experience. According to the work of vision scientists such as

Alva Noë and Kevin O'Regan, the visual world is not really a wraparound cinema screen; it's that web of expectations that link movements of my eyes and head to changing visual sensations.[5]

Detailed experiments have shown that what we actually 'see' in any moment is much narrower and sketchier than we think. For example, experiments on what are called 'change blindness' and 'inattentional blindness' show that, unless we are actively attending to a specific location, we are really poor at spotting even quite gross changes that occur there. People talking on their cell phones very often do not see the brightly clad clown on a unicycle who rides past them. Searching for a friend at a concert, we bias ourselves to look for flashes of yellow, because we know she is wearing a lemony shirt. So we notice lots of yellow things, and fail to recognise our friend when she walks by, having slipped on her coat.[6]

This way of looking at vision is a shock to the Cartesian view of perception. We naively think of vision as giving us largely unproblematic, objective access to the world around us, uncontaminated by considerations of subjective need. We see what's there, right, and *then* evaluate it and respond to it. We see 'seeing' as a receptive process, just noticing 'what's there'. But this way of looking at seeing obscures its deep relationship to doing and needing. To see, we need the world to rub against those specialised bits of skin called eyes – or our eye-skin to actively rub against the world, identifying its textures and edges.

It may be that we have become such predominantly visual creatures partly because we have (mis)interpreted vision in this Cartesian way. If we think of visual perception as decoupled from the other two aspects of the body, doing and needing, we are encouraged to view it as the closest of all our senses to that disembodied and dispassionate 'mind'. 'I see', we say, when we mean we understand. Research on vision in psychology and physiology vastly outweighs that on all the other senses put together – because of its Cartesian bias.

*****

What is going on in the rest of the body alters the way our skin behaves, so our 'somatosensory perception', as it is called, is constantly being tuned and primed by what is going on in the brain and the gut (for example). Just as the permeability of the membrane of an individual cell changes, depending on what the cell as a whole needs to absorb or expel, so our big 'membrane', the skin, alters its behaviour in the light of what is going on elsewhere. Though we may not be aware of it, the autonomic nervous system is constantly modulating the activity in our sweat glands, and this changes the electrical conductivity of the skin, formally called the electrodermal activity (EDA).[7] Stick two electrodes on your skin, and a meter will show the fluctuating level of the skin's ability to conduct electricity. The sweatier you are, the greater the conductance. Because our skin is hooked up with changes occurring all over the body, the EDA is often used as an indicator of our overall level of physiological arousal. We know that we may flush with fear or embarrassment, and our 'hair stands on end' when we are excited or enraged (as threatened cats' and dogs' hackles rise, too). But skin is involved in the life of the intellect as well as the emotions. When we face a tricky decision – as in deciding the answer to an item on an intelligence test, for instance – our skin is as involved as our brain. In fact Damasio and his colleagues have found that the EDA can be a more sensitive indicator of our thinking than our conscious minds are. Intelligence is a whole-body happening![8]

Emotions also change the colour of our skin, by the way, as blood is sent or withdrawn. We blush with shame or go pale with fear – so skin colour acts as a very important social signal. The other senses have the same capacity as the skin to act as social signals. The muscles of the nose and mouth signal disgust or amusement. We sniff in disapproval. Eye movements can give away our intentions: our pupils dilate when we are feeling sexy. According to folklore, even our ears go pink with pleasure and burn when we are being talked about.

All the senses show the same close association between perception and action. I smell mostly by sniffing. I get a whiff, and then I intensify my smelling by inhaling sharply. I taste by chewing and smacking my lips. I don't just hear, I listen by turning off the radio, becoming very still and cocking my head in the direction of the noise. Even when I am apparently still, perception is dynamic. Nose and mouth are full of specialised receptors that are literally touched by a wide range of molecules. The ears are designed to be touched by pulses of pressure in the air, just as our eyes are designed by evolution to be stroked by a spectrum of electromagnetic energies.

The fact that movement is integral to perception is starkly demonstrated by a famous experiment that every psychology undergraduate learns about.[9] Two kittens were yoked together in a kind of primitive two-seater carousel for the first few weeks of their lives. One had its feet on the ground, and could at least walk round in circles as it wished. The other was forced to lie in the cradle opposite, and be carried around at the whim of its more fortunate sister (see Figure 4). This devious device ensured that the two kittens had the same kind and amount of visual experience, but for one this was linked to its own movement, and for the other, it was not. Despite plenty to see, and despite no obvious damage to its visual system, the passive kitten never

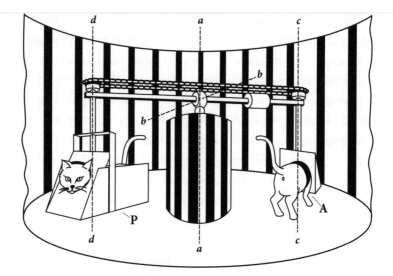

Fig. 4 The Held and Hein cradle. Both kittens get the same visual experience, but for only one of them is this linked to their own actions.

learned to see properly. It couldn't recognise objects, navigate its way around a room, or see in depth. Deprived of the opportunity to link movement and vision in a meaningful way – to discover how the world changed as *it* changed – the kitten never learned to see meaning in the world. If human babies are swaddled and bound so they cannot move, their perceptual and cognitive development quickly goes awry. Those poor orphan babies in Ceauşescu's Romania, stuck in cots for months on end, did recover somewhat, but never fully.

Needs, Deeds and See'ds

Our bodies are defined, broadly, by three sets of considerations. The first, and most basic, is: what do I need? At every moment, the body is alive with a variety of concerns and priorities, all of which have a degree of urgency. I want to get this chapter finished, and I need to get to the shops, and I have to break off at some point and deal with the important emails, and I have to remember to put out the bins ... Some of these concerns are physiological and some reflect longer-term desires, interests and obligations. Some spring from my beliefs and attachments and my self-image. Some of them are mutually compatible, but others compete or are even contradictory. (I want to see myself as a morally admirable being, for example, but I am late for the meeting so decide not to stop and help the person in distress – and feel guilty.) Some of these, like the need for air, stay pretty constant over my lifespan; others develop, and some peter out, along the way. Generally, though, in the course of growing up we develop a rather large and entangled range of priorities. They constitute a substantial, fluctuating, partly conflicting portfolio of concerns: all the things I have on my plate right now. Let's call these our Needs, for short.

Second, there is the collection of my reflexes, skills and capabilities: the know-how I possess, all the way from deeply embedded and unconscious regulation of my blood-sugar levels (assuming I am not a diagnosed diabetic) to the learned expertise that enables me to concoct a tasty meal out of leftovers, or indeed write a book. Again, some of these are deeply wired in and apparently beyond conscious control; others are more or less unconsciously picked up along the way as I incessantly tune my capacities to respond to the changing world; and some are acquired deliberately and often with a good deal of conscious effort and practice. I have the ability to carry on typing, or make a cup of coffee, or play a sliced backhand, or sing the Hallelujah Chorus, or ... or ... or ... These skills *in toto* constitute my portfolio of capabilities. For short, I'll call them my potential Deeds.

And finally there is, at every moment, a portfolio of opportunities being revealed by my senses. Let's call them See'ds. In my current world, as it appears to me, there is a television that could be switched on, a window that could be opened, a magazine that could be read, a keyboard that could be typed – but not a submarine periscope that could be lowered, or a glamorous film star yearning to be kissed. Perception's job is scoping out the possible 'theatre of action' – a sense of all the things that current circumstances permit me to do – so that I can select and craft my actions appropriately. The Yiddish word *klutz* describes a person whose actions constantly miss the mark because they misread the situation. (Think of Mr Bean, Lieutenant Frank Drebin or the magnificently incompetent Inspector Clouseau.)

Why you have a brain

At any moment I am a buzzing swarm of Needs, Deeds and See'ds. And my job – the function of my intelligence – is to resolve all this shimmering mish-mash of information into an answer to the perennial, deceptively simple-sounding question: 'What is the best thing to do next (all things considered)?' And for this I need a brain. An amoeba has few Needs, even fewer possible Deeds, and very limited See'ds, so it doesn't need a proper brain. But I do, because deciding on the best thing to do next is often, as my Oxford philosopher friends put it, a deeply non-trivial problem. Sometimes this multivariate equation is easy to solve: it is literally a no-brainer. When I am in the middle of a well-practised routine, like cleaning my teeth, a sequence of actions unfolds automatically (unless I am disturbed by an event or a thought). I have plenty of habits that tell me, 'When I need to do X, and the world is like Y, then do Z'. When I want to make a white sauce, and I am in my own kitchen, I go to the shelf by the door and get the flour.

But when I want to resolve an upsetting conversation with my wife, and she has gone off to work in a frosty mood, then I – my body-brain – may need to consider a wider and less routine array of options. I could send her a friendly and apologetic email and offer to take her out for dinner and try to work it out – but the outcome of this is far from certain. She may not yet be in a conciliatory mood. Even going to fetch the flour is thwarted if I have forgotten to replenish it. Every action I make is a 'best guess', based on expectations extracted from the past, which may, in the event, go wrong or prove inadequate. So there is always the need to reconsider, think on my feet, and recompute what 'the next-best thing to do' might be. That's where the intelligence of the body is most importantly to be found: in resolving unfamiliar combinations of Need, Deed and See'd into optimal, customised responses to novel situations.

This need to find an optimal (or at least adequate) resolution of three complex sets of factors is fundamental for all animals, so it is not surprising that the body-brain is evolutionarily

---

THE KEY QUESTIONS OF LIFE (AND DEATH)

WHAT DO I NEED?

WHAT ACTIONS COULD I PERFORM?

WHAT DO CIRCUMSTANCES ALLOW?

SO ... WHAT'S THE BEST THING TO DO NEXT?

---

designed to do it. And this means that, far from Needs (motivation), Deeds (action) and See'ds (perception) being three separate compartments, in need of being tied together by 'the mind', they are in fact knitted tightly together in the structure and functioning of the body-mind itself. Brains evolved to support their bodies in doing that knitting. Brains evolved to help increasingly complicated, mobile bodies deal with problems of coordination and communication that they could not solve on their own. It was – and remains – their *raison d'être*. Intelligence did not arrive from 'elsewhere', like a newly appointed Managing Director, proud of her Harvard MBA expertise, ready to 'kick ass and take names' in the corporation of which she was now the Boss. That intelligence pervades the body and its servant, the brain. The brain is the central information exchange of the body where these three swarms of factors can come together and, through communication, agree on a plan. The brain does not issue commands; it hosts conversations.

The interwoven brain

Needs, Deeds and See'ds are automatically and very nearly instantaneously integrated in the brain.[10] There are systems and pathways in the brain that link perception and action directly, so that, within a few hundredths of a second of my seeing you begin to reach out for the last piece of sushi, my own motor cortex is already beginning to construct a pre-emptive strike (which, in the interests of other considerations such as friendship, may, another few hundredths of a second later, be vetoed). At the same time as my brain begins to pull together the threads of the action, the motor cortex is telling the sensory cortex what changes in the world to expect as a consequence of my launching the action. The somatosensory cortex, the bit that registers changes to the body, is being primed to expect changes in the way my right arm feels, and the visual regions of the brain are anticipating seeing a hand appear, wielding a pair of chopsticks and moving rapidly towards the dish on the restaurant table between us. As a result of lots of previous actions, my movements are partially and automatically encoded in terms of their anticipated effects on perception.[11]

There are many demonstrations of this tight coupling of action and perception. Imagine that you are shown 2D pictures of two 3D objects and asked if one can be rotated so that it is identical to the other. Not only does the length of time to do this depend on the angle of rotation you have to turn the shape through; brain imaging shows that the motor areas of the brain are active while you are doing the task. You don't just watch the shape being spun round; it appears that your brain actually has to do the spinning. In another appealing demonstration, Rob Ellis and Mike Tucker at the University of Plymouth showed people a range of pictures one at a time, and they simply had to press one button if the picture was of a kind of water jug, and another button if it wasn't. All the jugs had handles, and they were photographed in profile so the handle was prominent – but the pictures varied in whether the handle was on the same side as the 'Yes' button or on the opposite side. People were much faster at correctly pressing Yes if the relevant button was on the same side as the handle! Seeing the picture automatically activated the appropriate hand movement for picking it up, and if this was compatible with pressing the button, the two forces combined to make people faster. If the two movements are incompatible, their conflicting tendencies slows us down. Seeing *is* doing, apparently.[12]

In practical terms, what you can do about something influences the way it shows up in your perception. Without touching them at all, you see objects and pictures that are placed near your hands differently from those that are further away. They look clearer, they are looked at longer, and it is harder to shift your attention away from them to another object. Richard Abrams and his colleagues at Washington University in St Louis speculate that this is because things closer to the hands are more likely to be things that are relevant to us – such as tools or food, for example – and so automatically take priority over other aspects of the world.[13]

Which brings us on to the embroiling of need in perception. Just as action and perception are tightly stitched together, so are they bound in with information concerning values, concerns and interests. Perception is not neutral: it is already weighted, with no conscious thought or awareness, by a host of motivational factors. Hills actually look steeper to people who are tired, ill, elderly or wearing heavy backpacks. The physical cost of climbing the hill is already factored in to the way the hill looks. People who are afraid of heights judge a balcony on which they are standing to be further from the ground than people who don't have that fear. And golfers who have just had a good round literally see the hole as bigger than when they have had an off day. Hoping, wanting and fearing are already dissolved in perception, in other words. We don't have to add them in deliberately.[14]

Our bodies understand ideas in terms of what they are good for, and how we make use of them: not just in terms of the features by which we recognise them. When you ask someone what a 'ball' is, in the context of a discussion about football, bits of their motor cortex light up that correspond to control of the legs and the act of kicking. Ask them the same question in the context of talking about tennis, and the embodied 'meaning' that is automatically activated involves arms and shoulders. If you had just been talking about your teenage daughter's school

Fig. 5 Seal/donkey illusion.

prom, 'ball' would activate responses to do with dancing (as well as visceral anxieties about the cost of the dress). When I hear 'John took the book' and 'John took the needle', the motor cortex primes itself to make two quite distinct kinds of grasping action – even though there is no question that I have actually been asked to 'take' anything literally myself.

Even social emotions such as embarrassment change the way we see the world. In a rather unkind experiment, Emily Balcetis and her colleagues made their volunteers play a game in which, if they were shown a picture of a farm animal they would then have to judge a singing competition, whereas if a picture of a marine animal came up, they would immediately have to do their own karaoke performance and be judged on it by others. On the critical trial, they were shown ambiguous pictures such as the one in Figure 5, which could be seen either as the head of a donkey or the body of a seal. People were much more likely to see – literally see – the version that got them off the hook of having to perform![15] This was not a matter of cunning thinking; it was the body-brain doing what it is designed to do – far faster than thought.

Information from your viscera, your muscles and your senses arrives in your brain through different gates, but within an instant the different types of information are chatting animatedly to each other. Especially as you get to know the world better, your brain circuitry remembers what sights and sounds went together, with what actions and reactions those sensations were associated, and what the felt consequences were. All this is filed away so that, next time, you will be able to integrate the various considerations in a smoother, faster and more satisfactory way. There is always the chance that my best-laid schemes will go awry but, on balance, my brain gets cumulatively better at resolving the issue of what 'the best thing to do next' should be.

One way of putting this is to say that our brains are profoundly egocentric. They are not designed to see, hear, smell, taste and touch things as they are in themselves, but as they are *in relation to our abilities and needs*. What I'm really perceiving is not a chair, but a 'for sitting'; not a ladder but a 'for climbing'; not a greenhouse but a 'for propagating'. Other animals, indeed other people, might extract from their worlds quite different sets of possibilities. To the cat, the chair is 'for sleeping', the ladder is 'for claw-sharpening' and the greenhouse is 'for mousing'. To me, my old university friends are 'for reminiscing and silly joke telling'; for some of our long-suffering partners, they might not be for anything much except perhaps 'for putting up with'. A block of wood that affords only 'burning' to me might well afford 'displaying' to an inveterate beachcomber, or 'carving' to a sculptor. My laptop affords 'reprogramming' to my IT-savvy friend Charlie, but only 'word-processing', 'emailing' and 'internet-searching' to me. To someone who is recovering from a stroke, the stairs in her house may no longer afford 'for ascending', and they will, in consequence, look different.

In fact, we could see the deep structure of the brain in terms not of Needs, Deeds and See'ds but in terms of a different set of basic concepts that have already combined these (see Figure 6). If we meld perceptions and actions – if we see perception as deeply imbued with the possibilities for action – we could call the result *affordances*. An affordance is a scene already parsed in terms of the things I could possibly do. If we blend perceptions with our concerns, we might call these *opportunities*. An opportunity is an aspect of the world seen in the light of my current needs, interests or values. And if we combine actions and concerns, we could speak of *intentions*. An intention is an incipient action that already has a sense not just of possibility but of purpose.

This flipping of the categories is like the transposition we can make of the primary colours. In primary school, we are told

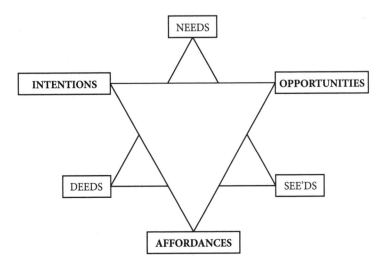

Fig. 6 The integration of Needs, Deeds and See'ds.

that the three basic colours are red, yellow and blue, but when you replace your printer cartridges, you will buy cyan, magenta and yellow. What look like secondary or combined colours from the primary school perspective can, for some purposes, be better seen as the basics. I think that may well also be true of the three 'primary' hues of the functional human body. It takes further work by a sophisticated nervous system to pull affordances, opportunities and intentions apart, and decouple perception, action and motivation from each other. To see a situation with judicial impartiality does not come naturally to us. It is a difficult cognitive trick, which takes years of schooling – studying things you do not care about and cannot make use of – to master.

Anticipation

If we waited till the world showed up, we would often be caught napping. By the time she can actually see the lion, it may be all over for the gazelle. If I wait till the guests arrive to remind me that I need to cook the meal, we will have a late dinner and disgruntled diners. So another key design feature of the brain is anticipation. If I can make a good guess as to what the world is about to do, I can prepare my response and mesh more smoothly and speedily. As neuroscientist Marcel Kinsbourne says, 'Anticipation is a wager based on previous experience. It readies a response to an event that has yet to occur.'[16] The world and I become, as much as we can, like familiar dancing partners: we can flow and improvise together because my deep bodily knowledge of you, and yours of me, enables us to 'read' the slightest shift in the pressure of a hand, or the angle of a hip, and begin to respond almost before the signalled movement has actually begun. That's why, as the great psychologist William James, observed, every moment of our conscious lives is not just 'here now', but is infused with a continuing sense of the impending and the receding. To coin a phrase, time past and time future are both contained in time present.

We store our rolling knowledge of the world in terms of such expectations. I know what is over there not because I can see it, but because I have a pretty good idea of how my sensory experience would change if I flicked my eyes and angled my head in that direction. I anticipate how my movements would influence my perceptions, so my construction of the world is a fine tissue not of actual scenes and objects but of these predictions. 'I believe there is a tree over there because I know how to make it present to my senses should I need to.'

How does prediction work? At its most basic, by what is called 'spreading activation'. The brain is just a giant tangle of interconnected wiring with activity (electrical, chemical and physical) running through it. Some of the joints in this tangle are worn by experience, so when a pulse of activity comes to a junction, it will, all other things being equal, take the path of least resistance. As trains of pulses travel around the network, they can change its behaviour in two ways. They can leave a lasting facilitation of the chosen pathway, so that future pulses are more likely to go that way. Or they can leave an amount of activation behind, so that the path is primed, but not irreversibly changed. Some of these dollops of priming fade quite fast, but others can be 'chronic', in that they can change the routing in a functional but not a structural way. Both structural and functional changes are mechanisms of prediction. Note, by the way, that the words anticipation, prediction and expectation refer, in this context, to the way the brain is working, and not to any kind of conscious experience. Many predictions bias the *behaviour* of the body-brain without themselves becoming conscious.

Prediction streamlines the process of perception itself. Several researchers have recently picked up on suggestions by the nineteenth-century German physicist and physician Hermann von Helmholtz and spun these into an intricate new theory called 'prediction coding'. Basically, the brain is constantly generating its best guess about what's out there, and then feeding predictions, based on that guess, out to the different sensory receptors to create a downward tide of centrally generated sensory expectations meeting an incoming tide of sensory information. When they meet, any signals that match a prediction are cancelled out: there is no need to forward that information further up the incoming chain. Only information that is not expected gets passed up to a higher level, where predictions are adjusted or a whole new set of expectations are activated that do better justice to the input, and they are propagated back down in another wave of anticipation.

Through as many cycles of this process as are necessary, and often all in a flash, the brain finally settles for a good-enough, though always provisional, hypothesis about 'what is out there' and how it meshes with my current portfolios of concerns and capabilities. And this hypothesis incorporates already half-constructed plans and programmes for action. ('Ah – there goes the baby monitor – just as I'm trying to play this tricky hand of bridge . . . what kind of cry is it?' I mouth 'sorry' to the rest of the table, mime requesting a pause in the bidding, then incline my head towards the speaker, hold my breath, and await the next cry . . . OK, maybe just a sleepy whimper . . . doesn't have that sharp edge it does when she's upset – I'll see if I can play this hand out and then go and check . . .)

Though this sounds complicated, it is actually highly efficient in terms of the 'bandwidth' of information-processing capacity that the system needs.[17] And there is a good deal of neurophysiological evidence to support it.[18] There are indeed massive outflows from the 'higher' processing centres in the brain down to 'low' levels of processing in the eyes, ears and skin. But what it means is that we never do perceive 'what's really out there'. Our perceptual world is always powerfully imbued with the knowledge, needs and capacities our body-brains bring to the situation.

Perception is a fabrication – a hallucination. But it is a hallucination that is constrained by the facts. It is put to the test of experience time and again, and if it works, it stays. And this is yet another kind of anticipation. If I act on the basis of the perceptual model – a tapestry woven out of threads of Need and Deed as well as See'd – does the world react in the way the model says it should? If, having won the hand, I find the baby lying in a funny position and making faint mewing sounds that I don't recognise, a variety of powerful reactions kick in – one of which is my brain urgently updating its web of interpretations and expectations. Next time the monitor goes, my wave of predictions (and bodily reactions) will be different.

Finally, there are the predictions of what the sensory consequences of making that action will be. If I throw the ball so, I should see a trajectory rather like this. If I let go of the bottle and I am standing on the stone flagstones in the kitchen, then I can anticipate the length of the delay before I hear the sound of breaking glass. When the motor cortex is planning a movement it sends a carbon copy of the instruction to the sensory cortex, effectively saying, 'I'm about to do this, and so, all being well, you should be about to experience that.' If I try to tickle myself, the effect is underwhelming – because my brain can tell what is coming and cancel it out. (The very first paper I ever published in psychology, back in 1975, was entitled 'Why can't we tickle ourselves?'.)[19]

This ability to anticipate how what I do changes what I experience gives us another way of streamlining our operations. As we saw above, what is predictable is often of less interest than what is not, so we can cut costs by skimming over the matches and concentrating on the surprises. This is vital, because it is from the unexpected that we learn the most, and it is also the unexpected – that for which we are unprepared – that can be dangerous or disruptive. In addition, this kind of prediction enables us to tell the difference between changes in the world that are brought about by our own actions, and therefore possibly under our control, and those that occur independently of what we are doing, so are beyond our immediate control. How we draw this distinction makes a big difference to how we treat events.

One of the benefits of being able to make this distinction is the fact that the visual world stays steady as we move our eyes about. As we look around, or as we read a page, our eyes are flicking about all over the place. The patterns that fall on the retina are changing continuously and grossly. Yet we experience the world as both stable and largely unchanging: a stationary backdrop against which unexpected movements (as well as disappointments) instantly stand out. This stability, in the face of so much change, is achieved by the brain using what it has just seen to anticipate the visual changes that its own eye movements are likely to make and cancelling them out of the visual equation. Close one eye, and nudge the outside corner of the other with a finger. You will see the world move. Even though it is your own finger that is causing the movement, your brain has not set up the same kind of cross-referencing between 'retinal displacement' and 'fingertip movement' that it has for 'direct brain-initiated eyeball movement', so the sensations are not cancelled out.

Where the correspondence between action and perception is very strong and familiar, the action itself can generate the

Fig. 7 A Kanitzsa illusory pyramid.

perception that normally goes with it, even when it isn't there. If you and a friend go into a pitch-black room and she waves her hand in front of you, you might feel the slight disturbance of the air but you won't see anything at all. But if you wave your own hand, you will have the powerful impression of being able to see your hand. Your brain fills in the gaps and you see what you would normally expect to see, in just the same way as the more familiar visual illusions, like the Kanitzsa figures (see Figure 7), replace what is actual with what is probable. You 'see' the illusory edges and depth because your brain assumes that the reason the black circles have bits missing is because there is something in front, and also that the different shades of grey reflect 3D shadows – so it helpfully doodles on reality to make it fit with that expectation.

*****

All told, the science of embodiment asks us to adjust our understanding of the brain in several ways. The brain by itself is not the physical substrate of 'mind': we can't just take all the clever things that we used to attribute to the ghostly mind and plop them into the brain, leaving the rest of the body to be – in George W. Bush's immortal coinage – misunderestimated in the same old ways. Body and brain function as a single unit. Though parts of the body can and do talk to each other directly, they also need to send missives and emissaries to the standing conference in the brain, where the really knotty problems and conflicts can get ironed out and prioritised.

Nor is the brain organised into a neat series of processing steps, like an old-fashioned production line, that lead from perception, through thought and memory, to action. Though different kinds of information do enter and exit through different doors, once they are through the doors what goes on in the brain is much more like an animated party than a game of Chinese Whispers, with Ears and Skin, Fingers and Shins, Lungs and Intestines all chatting to each other. There is no separate compartment called Memory; memories and expectations are the stuff of all of these conversations. And there is no Chief Executive who steps in to resolve disputes or correct impressions on the basis of her higher experience and intelligence.

These insights offer us a new logic of the brain, and a non-Cartesian way of thinking about its place in the body. But we can do better than that: we can delve into the real-time workings of the brain and the body, and see up close how some of these conversations happen, and where. That is the business of Chapter 5.

HOW BRAIN AND BODY TALK
TO EACH OTHER

*I used to think that the brain was the most wonderful organ in
my body. Then I realised who it was who was telling me this.*

Emo Philips

To understand how intricately the brain is hooked up to the
body, it is useful to have a rough overview of how the brain
itself is organised. Luckily, Don Tucker, a psychology professor
at the University of Oregon, has created just such a map of
the whole brain in terms of its core functions. The map is
three-dimensional. The dimensions are a bit rough and ready
but they will help us see the wood for the trees. The first dimension runs (very broadly) from the largely sensory back of the
brain (vision, hearing and touch are processed behind the
so-called central sulcus: smell and taste are processed deeper
down in the brain closer to the core) to the predominantly
action-oriented front. Sensory information is arriving in the
brain from many sources all at once, yet, on the behavioural
level at least, we can only do one thing at a time so there is a
progression along this dimension from multiple sources of

information working in parallel to a more sequential, one thing at a time, organisation.

The second dimension differentiates the more routine- and habit-based left hemisphere from the more diffuse and creative right hemisphere. We have two hemispheres, it has been suggested, so that we can run two modes of attention simultaneously: one that is focused and analytical, the other that is more synoptic and holistic. Both of these modes are very useful, and each hemisphere can vary its degree of focus somewhat. But it gives us a big processing advantage to be able to have both modes running at once, and not having to segue between them.[1]

The third dimension, however, is the really important one for our discussion here. It runs from the innermost core of the brain, the brain stem, upwards through the structures of the limbic system to the outer 'shell' of the brain, the neocortex. This core-to-shell dimension carries information about needs and concerns – matters of personal significance – from the signalling systems of the body up into the outer regions of the brain where it is progressively integrated with information arriving through the senses, and also with the developing organisation of appropriate motor responses. And then messages are sent in the reverse direction, back down the chain to the brain's core, whence the information can affect what is going on in the recesses of the body in the light of that centralised conversing and decision-making.

Moots and loops

Contrary to the Cartesian view, there is no big boss in the brain who forces through resolutions and dictates policy. According to the emerging perspectives of embodied cognition, the body is self-governing. It is like a medieval moot, a meeting that can reach a conclusion only by a process of respectful and attentive debate. A 'moot point' is one to which there is no easy or obvious answer and which therefore has to be referred to the moot. Much of the work of the body does have routine solutions, so no brain-based conversation is needed. But – especially in complex social worlds – moot points continually arise, and for these the central conclave is essential.

In Chapter 3 we saw some illustrations of how bits of the body talk directly to each other. But where information from muscles, guts, senses and glands has to be carefully coordinated in real time, the brain needs to be in the loop. And loops are what it needs. All the different bits of the body have information loops that run through the brain. The inward part of the loop from the ears or the stomach carries information about what is going on 'out there', and what needs or concerns there are 'down there'. The ears might be saying, 'Too loud, man' (you might remember Stan Freberg's version of the Banana Boat Song). The stomach might be saying, 'I could murder a cheese sandwich'. The lymph glands in the neck might be saying, 'Uh Oh, cold virus alert! Send us some more lymphocytes, willya?' And looping back out from the brain are complementary channels that carry advice, information and resources in response.[2]

The anatomy and physiology of these loops is incredibly complicated and intricate and I am not going to attempt to do anything more than provide a few illustrations here. To recap, there are three kinds of information travelling around your body, chemical, electrical and physical. Electrical impulses, like those of an old-fashioned telephone system, run via the fibres of the autonomic and central nervous systems to and from the brain. They are rapid and can be very specifically targeted. The chemical messengers flow through the bloodstream and the lymph system, constantly coursing through the body, and can alter the signals that the nerves are sending. The chemical systems are slower than the electrical, though blood can travel from the heart or the gut to the brain in a second or two, so even 'slower' can be pretty quick in real time. And obviously the chemical systems are less accurately targeted: molecules and microorganisms have to float around until they happen to find a 'lock' to which they have the right 'key'. This can be fine, if the distances the messages have to travel are small (perhaps from one side of a synapse to the other) or if a general broadcast is what is required (in an 'all hands on deck' kind of emergency). Having both forms of communication is clearly better than having either alone. In addition, many of these molecules are able to talk directly to the brain.

The third kind of whole-body communication, the physical, is often forgotten. But whole-body movements, as well as the continual ten-a-second waves of vibrations, carry information from here to there, and also help to coordinate activities in different areas of the body. For example, science confirms what we all know: that a walk after a large meal aids digestion. The gentle bumping and twisting of the whole body serves to massage and agitate the gut, which stimulates the release of digestive juices and stirs the contents of the stomach so that the juices can do their work more effectively. But movement plays a much more pervasive role than that. Coughing, sneezing, laughing and even yawning produce pressure waves in the body that (as well as squeezing the bladder) pump cerebrospinal fluid up to the brain, and increase blood flow, both of which change the concentrations of hormones and neurotransmitters that are swirling around the neurons and synapses. Orgasm and straining on the toilet may have analogous effects.[3]

Most organs, including the brain, are elastic and respond to whole-body movement with movements of their own. At the most gross level, a knockout blow to the head induces loss of consciousness, though why and how this happens is not yet known. At the micro level, every cell in the body is influenced by mechanical forces as the body and its different sub-systems twist and turn, tighten and relax, shake and tremble. Many functions of the cell – division, membrane permeability and gene expression, to name but three – are influenced by physical movement. In the brain, neurons physically twitch and turn when they are activated, and their axons propagate small pressure waves, as well as electrical and chemical ones. There are many protein molecules in the brain that both generate movement and respond to it. This small-scale physical activity may well affect the way messages are routed in the brain and the way neurons learn. A little is known about these mechanical effects, but this is a new field of research and a great deal remains to be discovered.[4]

Much of the chemical and physical information from the body is converted from changes in pressure, temperature, pulse rate or glucose concentration into electrical impulses within nerves, and these electrical messages, originating in the body, can simply be passed upwards, from synapse to synapse, to the brain. Pressure receptors in the walls of the arteries, for example, detect the physical ripples and distortions that occur with every heartbeat. The stretching stimulates electrical impulses that are carried up the vagus nerve and onwards into the brain stem, the midbrain and thence to the neocortex. Feedback loops respond to the signals of increasing blood pressure – in the healthy body – by sending back messages that slow the next heartbeat and open up the capillaries to allow more blood through, thus lowering blood pressure and keeping it within safe bounds. Only a tiny fraction of this kind of activity is accompanied by any sort of awareness, and so we tend to neglect its significance – though it underpins much of the somatic intelligence on which we rely. That's why chronically raised blood pressure is called 'the silent killer'.

So information from muscles, bones, organs and skin travels via the brain stem to the 'core' of the brain, and then onwards to the outer layers of the neocortex. The loops pass through a series of way stations where their signals are processed and blended in different ways to extract different aspects of the information they are carrying. The deep core of the brain comprises a vast network of such processing plants, many of which have exotic, bewildering or rather appealing names. There is the *parabrachial nucleus* (with many arms like an octopus?), the *nucleus of the solitary tract* (a bit of a hermit?), the *locus coeruleus* and the *nucleus accumbens* (which sounds as if it is in need of a lie down). A particular favourite of mine is the *periaqueductal gray*, which I think ought to be an undistinguished water bird. Antonio Damasio is particularly keen on the *superior colliculus* (a rather self-satisfied small rodent, perhaps?). Then on they go to more familiar parts of the brain such as the *amygdala* and the *hypothalamus*. As the journey proceeds, so specific information originating from different parts of the body is combined to generate a more comprehensive picture.

Chemical communication

A good deal of this somatic information, however, is not converted into electrical impulses, but communicates directly with the brain. Despite the existence of the blood-brain barrier, which prevents many hostile bacteria and toxins from entering the brain directly, there are many areas in the brain that can and do pick up and respond to changes in the nature and concentration of the molecules floating by. Regions of the hypothalamus and the hippocampus, as well as the neocortex, respond directly to the concentrations of insulin and other insulin-like peptide molecules in their vicinity, for example, and this may change the neurotransmitters, such as dopamine, serotonin and noradrenaline, that are being released into different areas of the brain.

The insulin receptors are especially involved in regulating appetite, and in allocating energy resources around the brain. We know that changes to bodily concentrations of insulin will lead you to eat more or less, or to select different foods. More surprisingly, due to its direct action on the brain, insulin can also reduce your inclination to persist with difficult tasks. Staying focused on something you find hard and would rather not be doing is energetically expensive, so changes to insulin levels may have a direct impact on your willpower. This is worth emphasising, as it is a very clear illustration of the relationship between body and intelligence.

Other chemicals associated with the immune system are also intimately and continuously involved in allocating and distributing energy around the brain. They influence which bits of the brain get the most 'juice', and thus determine, amongst other things, how efficiently the synapses will work, and the rate at which connections are formed. When energy is low, thinking is harder and learning is slower. We all know that we shelve difficult tasks when we are under the weather, and just want to watch television or read magazines. But these influences of body on mind don't just occur when we are ill; they are at it all the time.[5]

Our guts also talk to our brains via a host of chemicals. Peptides and hormones may be released from the walls of the gut in response to certain foods or the detection of toxins, and these, like insulin, travel to the brain and influence what it is doing. All you may be aware of is the desire for something sweet, or the need to visit the bathroom, but these conscious feelings and promptings are merely the gross result of a thousand small changes across the brain.

The gut is home to a collection of tiny creatures, the *microbiota*, so vast and so varied that they make Noah's Ark look like a suburban bungalow. In your stomach, right now, there are ten times more microorganisms than there are cells in the whole of your body. Mostly we live in peace with this mass of tiny tenants, though, as we know, they sometimes get out of hand, and what is going on with their giant host can also upset them. Stress at any age, but especially early in life, may throw this symbiotic relationship out of kilter and stop the immune system working as well as it should, for example.

Yet we cannot function well without the microbiota. Their waste contains molecules (short-chain fatty acids) that are essential for our health. Animals that are raised in a totally sterile environment, where there is no chance for bacteria to colonise the gut, react abnormally to stress. Worse, germ-free mice have decreased levels of a hormone that is vital for the proper growth of nerves, for example in the hippocampus, and this leads to deficits in their ability to learn and to solve problems. More amazing still, germ-free mice that were colonised with microbiota from a different strain of mice started to behave more like the donor mice than like normal mice with their own DNA! If we are anything like the mice, we are biologically designed to live in harmony with our bacteria, and they shape our behaviour beneficially. So we should be careful with the antibiotics and the cleaning products.[6]

We should also be careful about how we are born. In a normal vaginal birth, bacteria begin to colonise the gut of the baby as it is travelling down the lower birth canal. Babies born by Caesarean section, however, miss out on this colonisation, and this has implications for their brain development.

Caesarean-born babies have electrical brain activity that is less complex than babies delivered naturally. And rat pups born by Caesarean section show adverse differences, in the ways their brains develop, from those born normally.[7]

Small free-floating cells such as microglia and astrocytes are scattered throughout the brain, and these too are sensitive to changes in the concentrations of hormones and peptide molecules. If there is inflammation somewhere in the body, the brain quickly knows about it and responds, perhaps with a closing-down of perceptual interest in the outside world. Microglia and astrocytes may well be involved in altering the plasticity of the neurons, and in controlling the release and mopping up of neurotransmitters; they thus change our learning. Activated microglia release specific hormones that change the behaviour of other microglia across wide areas of the brain, thus magnifying any initial effect.[8] In turn, organs like the pineal and pituitary glands can be instructed by the brain to secrete chemical messengers that will travel back to the provinces and regulate their activity.

All this research confirms the view that the brain is not just in the business of telling the body what to do. Bodily activity is influencing brain activity just as much as the other way round. Body and brain are tied together so intricately and so rapidly that it makes no sense to locate all the 'intelligence' in one and none in the other.

How the brain maps the body

Out of this maelstrom of physio-electro-chemical activity the brain creates a series of complementary maps of what is going on in the body. A good map is a useful distortion of reality; it picks out what is valuable, for a particular purpose, and chucks away the rest. The map of the London Underground is a really successful map precisely because it ignores almost everything about London, and misrepresents the remainder. (The distance between two stations bears scant relationship to their distance apart on the map.) So the brain extracts useful maps from the tumult of information it is being sent by every other part of the body: maps that can be read alongside each other to tell us what the best thing to do next might be.[9]

The nuclei in the lower and middle parts of the brain start out by mapping quite specific features of the body. Some make maps of the current position and state of tension of the limbs, or of the overall state of physical balance of the body. Some distil information about the state of the body's different needs, for example on glucose levels, blood oxygenation, temperature and lactate levels in the muscles. Some record injuries and illnesses and draw out information that may surface into consciousness, after the maps have been redrawn a dozen times, as feelings of aches and pains or nausea. And some extract descriptions of the different kinds of touch that are occurring at the surface of the body: touches to the skin, to the lining of the nose, to the taste buds, to the tiny bones and muscles of the ears and the photosensitive skin, the retina, at the back of each eye. Temperature readings from various bodily locations can be summed and averaged to create an overall reading on the inner thermometer that triggers sweating or shivering, or makes us put on a sweater or head to the fridge for a cold beer.

By turning temperature and blood sugar and muscle tension into the common language of neural signals, different modalities can be integrated with each other to achieve higher-order mappings. The superior colliculus, for example, receives maps from the eyes, ears and skin, and begins to combine them into multimodal representations of objects in space.[10] Other regions start to combine different sources within the body. Localised activities in the immune system – dealing with a paper cut to a finger, say, and feeling queasy from last night's seafood buffet – get passed along and aggregated so that decisions of priority can be made. The sting of the cut may recede into the background as the feeling of nausea gets stronger. And finally, at the top of the loop, the portfolios of perceived opportunities, available actions and visceral concerns are mapped and put together, so that, if the phone rings or you are engrossed in a thrilling story, the ache in your backside from sitting so long on a hard chair goes unnoticed; or the stomach ache which had seemed so bad at breakfast magically dies down when you are told you don't have to go to school.[11]

Where it all comes together

Though these loops and maps are widely distributed round the brain and the body as a whole, there obviously needs to be a place where the high-level representatives of the body's various systems, their 'ambassadors', so to say, can confer. And there are indeed two structures that play a pivotal role in integrating these sources of information: the *insula* and the *cingulate cortex*. The cingulate wraps around the large bundle of fibres called the *corpus callosum* that connects the two hemispheres of the brain and so is ideally placed to collect and redistribute information from a wide range of sources. (*Cingulum* means *belt* in Latin.)

The front part of the cingulate has been shown to have a special role in evaluating how what is happening at a sensory level relates to the signals of need or value coming up from the body, and especially in detecting when things are not going well. It is then involved in talking to the action planning parts of the brain, in order to adjust or design behaviour that might be both advantageous and appropriate. This bit of the brain has been called the 'visceromotor cortex' to emphasise its role in integrating perception with both the visceral needs of the body and the motor capacities which it uses to respond.[12]

The insula – which should really be called the peninsula, as it is not quite an 'island' – is connected to many parts of the brain, including the anterior part of the cingulate and areas of the pre-frontal cortex, by a host of functional causeways. It is about the size and shape of a prune, and each hemisphere of the brain has its own insula. Bud Craig, a researcher at the Barrow Neurological Institute in Phoenix, Arizona, has a good deal of evidence to back his claim that the posterior part of the insula creates an overall sense of how things are in the visceral parts of the body.[13] Then, as patterns of neural firing are passed forward towards the front of the insula, this 'primary interoceptive representation', as Craig calls it, is integrated with other patterns coming from the muscles ('proprioception'), from the predictions and expectations alive in the higher centres of the neocortex, and eventually from the outside world. All this culminates, in the anterior insula, in what he calls a 'global emotional moment': the composite background feeling of 'how I am, right now'. It is the integrated sense which we refer to when someone says, 'How are you?' and we answer 'lousy', 'a bit off colour', 'so-so', 'pretty chipper' or, as my New Zealander friends are prone to say when feeling particularly well, 'like a box of fluffy ducks, mate'.[14] Psychology has long known that these moments integrate influences across a period of around a tenth of a second.

*****

Craig talks of these moments as if they were separate, like the frames of a film, but I prefer the image of a wave rolling through the ocean. On the gross level, the wave keeps its form, but the water – the content – keeps changing. At any 'moment' the wave represents the sum total of all the currents, swells and winds that are acting at that location. They come together to create a particular wave-form, with its signature composition and direction. Waves have a width; that is, they integrate the forces acting not at a point but over a small region of the ocean. The biological constraints on this span of integration might well account for the tenth-of-a-second duration of these apparent 'moments'. But each momentary wave is not separated from the moment before and the moment after. Like a real wave, it has both a leading and a trailing edge. It is simultaneously rising, existing and fading. In the rising are expectations and predictions of what the future may bring, and in the fading are echoes of the confirmations and surprises that arose from moments that have just been. (So the fading edge is where learning happens.)

(We might, if we were feeling whimsical, see the properties of the seawater itself as the capabilities of the body to behave; the currents and groundswells as the values and concerns in play at that location in time and space; and the winds as the influences from the external world. Even more fancifully, we could imagine a wave on a phosphorescent sea at night, in which the glowing white crest of the wave, standing out against the sea, represents our conscious awareness. The tenth-of-a-second integrations might then correspond to the way this ocean would look if illuminated by a strobe. More on this line of thought in Chapter 6.)

We should remember that the sensory and motor maps are also interweaving, as information goes higher up in the brain, to create other maps that already imbue perception with the

possibilities for action, and actions with the sensory consequences (marked as desirable or not) that might accompany or
follow them. Sandra and Matthew Blakeslee summarise the
research that shows how different areas of the brain cooperate
in this way:

> The sensory maps of your parietal lobe are also *de facto* motor
> centres, with massive direct interlinkage to the frontal motor
> system. They don't simply pass information to the motor
> system, they participate directly in action. They actively
> transform vision, sound, touch, balance, and other sensory
> information into motor intentions and actual movements.
> And by the same token, the maps of the motor system play a
> fundamental role in interpreting the sensations from your
> body. Your parietal lobe is not purely sensory, and your
> frontal lobe is not purely motor. *Physical sensation and action
> are best seen as a single sense that, like a coin, has two insepa-
> rable faces with different appearances.*[15]

(emphasis added)

The work of the insula is all in the service of better – more
fitting and effective – action. The more inclusive and well integrated is the information about current concerns, skilled capabilities and possible courses of action available in the world, the
better the selection and design of behaviour will potentially be.
Craig puts it more formally: 'Cortical integration of high-
resolution information on the state of the body provides
improved conditions [to] guide behaviour ever more efficiently.'[16] He argues that there is a clear evolutionary parallel,
seen in different animals, between improved ability to select
and control fluid responses and the development of this 'high-
resolution' analysis of bodily states and needs in the insula. The

anterior insula and the anterior cingulate are intimately interconnected, and almost always active at the same time. Together
at the place where the various loops of information come
together, the insula takes the lead on defining 'what's so', and
the cingulate leads on designing 'what to do about what's so'.

*****

Two further wrinkles to add before we leave the loops and
maps.

First, you'll recall that we touched earlier on the idea of
'predictive coding'. The brain doesn't create a replica image – a
'Google Earth', if you like – of everything that is going on in the
world. That would be cumbersome and unnecessary. It creates
guesses about what's out there and only adjusts them if they are
wrong. The same thinking has recently been applied to the
brain's representation of the body by Anil Seth, Hugo Critchley
and their colleagues at the University of Sussex. The brain's
high-level maps generate guesses – based on all the expectations they have built up – about the kind of feeling we are
having, what its significance might be, what might have caused
it, and what we might be about to do about it. 'I didn't sleep well
last night. I must be tired. So that feeling of ennui I've got (and
those yawns I'm trying to stifle) are probably just exhaustion,
not boredom . . .' This guess is then relayed back into the body
to see if it matches what is actually happening 'down there' (and
to send back mismatch messages if it doesn't). Such reasoning is
not conscious or deliberate, of course – or not usually – but it is
always part of the process whereby we come to experience what
we are feeling.[17] The brain does not need to build a replica
'body' inside itself – because it simply remembers where to go
and look in the 'real' body if it needs to. Roboticist Rodney

Brooks famously said, 'The world is its own best model'.[18] The same is true of the body. You only need to create models and maps for more subtle, distilled things that would not be self-evident in the body itself.[19]

The second wrinkle links to this. It comes from a suggestion by Damasio that the brain can create mini-loops (which he calls 'as-if loops') within itself to simulate the (wanted or unwanted) bodily effects of different perceptions and/or actions. Using the same high-level maps, the brain can run a simulation of what would (probably) happen in the body, if I did such-and-such, without actually having to engage the body directly – which would be more energetically expensive, as well as time-consuming. The brain takes its own guesses as to what is going on in the body (and what is causing it) for granted, and then proceeds to design actions on the basis of those conjectures. As I say, this saves time and energy – but it runs the risk that you may have got yourself wrong, in which case you will keep reacting to the present as if it were a rerun of the past when it isn't.

Two ways of moving; two ways of perceiving

Our actions in the world are, broadly, of two kinds. There is action *on* the world, when we are at close quarters. I might pinch it, unwrap it, kick it, play it, fight it, drink from it, type on it or paint it. For this type of action, my body is the origin, the *subject* so to speak, and action emanates from this centre, so I need an *egocentric* map. From this perspective, 'objects' – such as an 'audience' or a 'leg of lamb' – are seen in terms of their affordances for beneficial manipulation. How I see them depends on what I am up to. In this frame, other things and other people are, as Hermann Hesse once put it, 'cloudy mirrors of my own desire'.[20]

There is the tight knitting together of perception, capability and my agenda that we have explored so far.

But there is also action that enables me to move *through* the world so I can get where I want or need to be. In this frame, my body itself is an 'object' that travels through space, so I need a map in which I can locate myself as one object amongst an array of others. This is called an *allocentric* map. This map is more 'objective': it is like a chessboard on which I am the White King surrounded by other pieces, some benign and some hostile. The other elements in this space still have relevance to me, but I am at one remove, and so more inclined to see in them a range of potential affordances rather than only the one that happens to be at the top of my list when I bump into them.

These two kinds of action – call them manipulation and locomotion – require us to operate in two rather different kinds of world, so it is no surprise that the brain has indeed evolved two such complementary systems. These have been well explored, especially in the context of vision – the sense, par excellence, that contributes to the feeling of being an entity travelling through space.

These two systems can be persuaded to reveal themselves, and their different priorities, with the aid of some simple visual illusions. In Figure 8, the central circle on the left definitely looks bigger than the one on the right, and no amount of measuring (which shows they are actually the same size) will change the way you consciously see it. But supposing these patterns are laid out as coins on the table in front of you, and you are asked to grab the middle ones: do your hands make the same mistake as your eyes and prepare a grip that is either too big or too small to grasp the coin? No, they don't. If you are videoed as you reach out for the coins, and the recordings carefully analysed, you can see that your hand makes exactly the right grip to pick up the coins. The

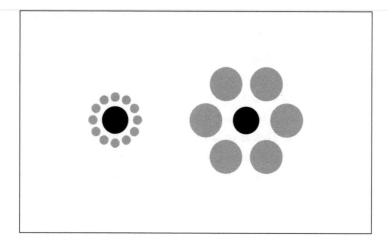

Fig. 8 Titchener circles illusion.

bit of the brain that is more concerned with immediate action than elaborated interpretation isn't fooled by the illusion at all.[21]

The locomotion system, by contrast, takes, as I have suggested, a more objective, general-purpose view of things. It tries to detach its description of the world from immediate perspectives and give a more neutral view of what and where things are and how they might turn out. This Big Picture perspective often adds in assumptions that are generally true in 3D space. The figure on the left, stylised though it is, reminds the brain of looking down on top of a scene, in which the ring in the middle is nearer to you than the peripheral rings. The figure on the right looks to the locomotor system a bit like looking down a hole, in which the ring in the middle, if it were 3D, would be 'further away' and therefore looks smaller. The more 'objective' locomotor system can be tricked by the illusion, but the manipulation system doesn't fall for it. If you are going to grab the central ring, you want to know how big it really is, not how big it is relative to its surroundings.

The manipulation system is evolutionarily older, and faster acting, than the locomotor system. (Think of a predator stealthily manoeuvring into position, and then pouncing.) It seems to be primed to make a quick assessment of 'what's out there' based on size, shape and distance – the things you need to know if you are to make an effective grab for it, or a quick gesture of self-protection – and also a first guess at its desirability: is it worth grabbing for, or defending against? Moshe Bar says, 'Like a Dutch artist from the sixteenth or seventeenth century, the brain uses … visual information to produce a rough sketch, and then begins to fill in the details using information from memory.'[22] This initial sensory sketch flashes across the top surface of the brain, via what is called the dorsal route, and activates information in the orbito-frontal cortex, an area that is known to weigh up the value, relevance and utility of experiences. In the case of seeing a face, for example, the brain's first sketch can tell the difference between a face and a non-face, and, crucially, can react differently to a happy as opposed to a sad face (and, a split second after that, breathing, heart rate and digestive processes are all beginning to be altered as well). Here we have another example of the way in which perception is almost instantaneously imbued with its relationship to our visceral concerns.

Cognitive control

As your body-brain is weighing up the situation it may well identify not just one but several courses of action that could be 'the best thing to do next'. Many of your sub-systems are clamouring for attention, but your body has to act as a whole. The body is like a potentially unruly choir rehearsing for a concert. The choir is a cooperative, so everyone can have their say, but

97

unless at some point we all agree on what we are going to sing, and in what order, the concert will be a shambles. We need some conducting – but the conductor is not a dictator, she is a coordinator, hired by the choir to help keep itself in line. And, at the same time as priority decisions are being made, the situation is often changing, so that you have to keep deciding whether to stick with the priorities you had a few seconds ago, or shift your focus of concern and explore new opportunities that may have just popped up. (New sheet music is constantly being published, so the potential repertoire is expanding.) Should you lock on to this, or take a peek at that? Your brain helps the body to synthesise, prioritise and sequence all these concerns and possibilities, and keep them under review. How does it do it?

Each of the possible courses of action under consideration gets tagged with a number of indicators that will help the brain to do its job of conducting the somatic orchestra. How *important* is it to achieve the goal of that action? How *urgent* is it? How much *cost* or *effort* will be involved in executing the plan? How *risky* is it: what could we lose if the plan fails or goes wrong? How *likely* is it to go wrong: how sure are we that we can 'pull it off'? And how sure are we that the information on which the plan is based is *reliable*? These tags are what Antonio Damasio calls 'somatic markers', because they rely on information in and from the body. They involve estimations of value, significance, apprehension, confidence and risk, which are all emotional issues. And they are based on past experiences with similar actions in similar situations: experiences of relief, delight, doubt or disappointment. When an action (or an element of an action) becomes a live candidate in a new situation its somatic tags are activated, and they are looped through specialised parts of the pre-frontal cortex.

These are designed to weigh up competing or conflicting claims of importance or legitimacy by a range of candidates, decide which is going to be the one that gets to steer the whole bus (for a while) – and stop the unsuccessful candidates from interfering!

Of course, there is no 'little person in the brain' who is making these decisions; the brain has to be organised so that a front-runner emerges automatically just by the application of some rules of thumb embodied in the way the neural networks function. These decision rules will determine how the activity in the different loops is added up and cancelled out. Whole books are written about how, in detail, this might be done, and which bits of the frontal lobes, coupled to which other bits of the brain, might be responsible for which operations. All I can do here is try to give a very rough sketch of the logic.

Remember that three sets of factors, each of them potentially complicated, have to be resolved together: the Needs, Deeds and See'ds. Sometimes the solution to this multivariate equation is clear-cut. There is an overwhelmingly urgent and important need ('Stop the car, I'm going to be sick' easily trumps 'But we're already half an hour late for dinner'). Or there is only one course of action open to me ('No, Mr Bond; I expect you to die . . .'). But often the competition is more of a close-run thing, and then the relative strengths of the tags have to be taken into account and the knockout rounds begin. Important but not so urgent? You can go on the standby list. Important but uncertain of success, or too costly? Sorry, we can't risk it at the moment.

The process is largely accomplished through the frontal lobes' ability to deploy highly targeted bursts of neural inhibition across the rest of the brain. Especially important here are the dorsolateral areas of the prefrontal cortex – the top and side-facing portions of the frontal lobes on both sides of the brain. If one goal is becoming pre-eminent, the competing goals get actively suppressed and I experience a visceral clarity of purpose (a burning desire). If the choir decides that we need to stick with a single programme for the time being, attention to all those exciting new possibilities is dampened down and I become impervious to distractions. These shifting, strategic patterns of inhibition make some concerns stand in line and allow others to go ahead. They manage attention so that it becomes focused and selective (as opposed to broad, inclusive or distractible). They turn some sources of information up and others down (so we can swivel the spotlight of awareness from the rumblings in the stomach to the sound of the birds, and back to the words on the screen).

When the choice between candidates for action is tight, or when a good deal hangs on getting it right, inhibition can arrest the selection process so that potential consequences can be checked out more fully and possibly revised or relegated, and thus our responses appear more slowly or more 'deliberately'. When these slower processes kick in, psychologists (who generally love to create binary oppositions) have often interpreted this as reflecting the activation of a different mental system, sometimes ingeniously referred to as 'System Two', that intervenes and restrains 'System One', a faster and more intuitive set of processes. But there is no need to parcel these processes up like this. Such bifurcations usually create more theoretical trouble than they resolve. I think it is only necessary to say that, under some (perceived) circumstances, the self-organising body-brain slows itself down.

There is an important corollary of this retardation. When inhibition stops an action flowing down into the muscles, or out into the lips and throat, it turns an overt action into an *imagined* action, or an actual utterance into a covert *thought*.

And this, as we shall see in Chapter 8, creates the breeding conditions for both consciousness and creativity.[23]

*****

So where is your brain? It is all over your body. And what is your brain? It is currents of information from all corners of your body continually making their way, by electrical, chemical or physical means, towards the stem of the brain, much of it by way of the spinal cord and the fluid canals of lymph and blood. Already subjected to some integration and simplification, each stream then moves up through various levels of the brain, being integrated with other streams of somatic information and transformed into a range of complementary 'maps' of the body's states of readiness and need. As these high-level maps arrive at the neocortex they are further integrated with information arriving from all the special senses about the state of the *outside* world, and with more information about what types of action might be available and appropriate. In the anterior insula, the dorsolateral prefrontal cortex, the anterior cingulate and the premotor cortex, processes are applied to these representations that sort out competing priorities and decide on which action plans will go forward for full-scale implementation. If the stakes are high, other frontal lobe processes monitor the implementation to see if any unexpected glitches or unintended consequences are appearing, ready to slam on the brakes and 'think' again.

The job of deciding what to do next is often very complicated – but it is made somewhat easier by the installation in the body of some pre-set modes of reaction to particularly significant kinds of event. Installed by evolution, these various modes are what we now call our emotions. It is to these that we turn next.

6

EMOTIONS AND FEELINGS

Emotions are . . . the result of the organism's need to continu-
ally monitor how things are going, and to initiate changes
within itself in response to possibilities for perceived harm
and benefit.

Mark Johnson[1]

The most obvious place where we experience our bodies, brains
and minds coming together is in our emotional lives. Emotions
involve muscles and glands, blood, sweat and tears, as well as
thoughts, memories and imaginings. An animated conversation may make me think, but it may also make me 'hot under
the collar', or give me 'butterflies in my stomach'. So, as we
move from the unconscious minutiae of neural networks and
biochemical soup to the world of daily experience – of
'mind' – it is our feelings and emotions that must be our next
port of call.

Before we get to the conscious aspect of our emotional lives,
however, we need to get a handle on what emotions are for.
Why do we get upset, angry and afraid? What is the point of

love, guilt or embarrassment? From an evolutionary point of view, emotions must be useful, or we would not have them. We must begin by seeing emotions as contributing to our ability to act intelligently, not as impediments to such action. To be sure, our feelings seem sometimes to 'get the better of us', and to be associated with courses of action that we felt unable to resist, but later regret. There are times when we wish we had gone with our head rather than our heart. But reason and emotion are lifelong partners who occasionally tread on each other's toes, not sworn enemies.

From classical Greece onwards, emotions like desire and anger have been treated as 'lawless wild beasts', as Plato put it, in need of strong restraint by the 'higher faculties'. Plato depicted the human personality as a chariot with two horses, one black and one white, which the charioteer continually struggled to harmonise. The white horse represented the virtues, and the black one (of course) was the wild and wayward stallion of the animal passions – lust and rage, basically – that was always trying to kick over the traces. It was only because of the chari-oteer's constant vigilance and stern discipline, and the exem-plary behaviour of the white horse, that havoc did not continually break out. Sounds familiar? Just over two millennia later, Sigmund Freud renamed the white horse Superego, the black one Id, and the charioteer Ego. *Plus ça change, plus c'est la même chose.*

This model of subversion, antagonism and domination even found its way into popular neuroscience for a while. A 1960s model of the 'triune' brain divided it into three layers: the 'reptilian brain' that comprised the basic life-support systems, the 'limbic system' that dealt with the emotions, and then the 'neocortex', or the rational brain, that did all the clever and honourable things and kept its two more primitive partners in check. When people got upset or afraid or angry it represented the brain 'downshifting' from the rational to the emotional, and our better nature struggled to shift it back up again. The job of the neocortex was to govern the rest of the brain, just as the job of the 'mind' was to govern the body.

Recent attempts to rehabilitate emotions through the popular idea of 'emotional intelligence' (EI) often miss the deep synergy between reason and emotion that I will try to demonstrate in this chapter. Either emotional intelligence is presented as if it were an intelligence *about* emotion, involving skilful control of emotion through techniques such as 'anger management'; or EI is seen as a kind of intelligence *of* emotion that is separate from other kinds of intelligence such as rational, linguistic or even 'bodily-kinaesthetic' intelligence.[2] My argument for the intelligence of emotion is different from, and stronger than, this latter idea. I want to insist that emotions are a deep, bodily-based constituent of every kind of human intelligence. Emotions are what make the world meaningful. If we perceived the world only with rational understanding, leached of emotional significance, we would not last long, and while we did last, we would find no fulfilment in our survival.

What emotions are for

The general view now is that our basic emotions are built-in 'default settings' of our whole embodied Systems. We feel emotional when we perceive something as belonging to a class of events that has a characteristic kind of significance. Mostly this means things that fulfil an important need, or threaten an important aspect of our well-being. Emotions are responses, developed through evolution, to important, perhaps even archetypal, events. If we see something as an object of desire we are automatically primed to approach and secure it. If we are hungry we start looking around for sources of food, approach them, and ingest them. If it is the mating season and love is in the air, animals of all kinds become sensitive to any sights, sounds or smells that signal the proximity of a possible mate. Conversely, if we see an aspect of our world as dangerous, we prepare to avoid or neutralise it. If a large shadow, typical of a predator, falls across an animal's retina it will engage certain actions (running for cover; exuding poisonous slime through the pores of its skin) designed to prevent it becoming someone else's lunch. Though they come in many shades, emotions are intelligent responses to events that are relevant to what we value – and what we value has its roots firmly in the physical body.

As we saw earlier, survival and well-being are better protected if we respond not just to significant events themselves but to signals and portents of those events. You are more likely to survive if you get ready to run when you see the grass twitch than if you wait to feel the hot breath of the tiger on your neck. We don't wait till we are in the school principal's office before we start to get nervous. Emotional states are often responses to those cues. They are not actions in themselves, but states of readiness to respond to events that, we suspect, might be about to unfold. We are on a hair trigger, so that, with the slightest confirmation that what we suspected is actually about to happen, we can leap into action and – sometimes, at least – beat the world to the draw. This built-in emotional intelligence adds a behavioural layer to the internal life support systems that we looked at earlier.

When one of these prototypical situations arises (gradually in cases like hunger; abruptly in cases like the sudden appearance of a threat), it is obviously vital that all the systems of the body react as one. Our Sub-Systems, often going about their business semi-autonomously, need to behave like volunteer firefighters when the siren goes. They need to drop what they are doing, take note of the kind of alarm that is going off, and start making the contribution to the general good for which they have been trained. Emotions enable us to switch a wide range of our faculties rapidly into the right general mode or stance with which to face the trouble (or potential benefit) that (we suspect) is in the offing. They are like the 'pre-settings' that come with the audio amplifier on a modern TV. A single touch on the remote control and you can change a whole variety of electronic parameters into a pattern that is optimal for 'drama' or 'classical music' or 'easy listening'. Each of the main emotions sets those values into a characteristically different profile, as we shall see. Evolution has equipped different animals, depending on the portfolio of challenges their habitat regularly throws at them, with a custom-built console of these pre-set buttons.

These whole 'body+brain+sensors' reactions are so intricately interwoven that it is impossible for us to pull them apart and tell what is 'cause' and what is 'effect'. The circular loops connecting body and brain are bi-directional, so that 'higher' processes are influencing 'lower' ones, at the same time as the 'lower' are feeding information up to the 'higher'. Words like 'resonance' and 'reverberation' capture this shimmering complexity much better than ideas of 'stimulus' and 'response'. Did I see the bear, feel afraid, and so tell my legs to run? Or did the unfolding of bear-seeing, gut-trembling and leg-thrusting happen in such a fast and loopy fashion that they are, is essence, different facets of the same holistic episode? Traditional 'folk psychology' tells us the former. Embodied science tells us the latter.[3]

Which bits of our bodies do emotions engage? The broad answer is: almost all of them. But it may help to highlight some of the most important. First, our internal physiology can be altered. As we saw in the previous chapters, this can include changes to heart rate and blood pressure, rate and depth of breathing, the physical and chemical behaviour of the intestines, and the chemical composition of the blood and lymph. In addition, aspects of bodily posture, facial expression and voice quality can change. Our shoulders may drop, our faces become angular and forbidding or tender and loving, and our voices grow hard or soft, loud or quiet. The skin changes colour as its blood supply increases or decreases – our cheeks burn with embarrassment or become pale with rage – as well as changing in sensitivity and sweatiness, and our body hair can stand on end. The big action muscles of arms, legs, shoulders, neck and so on can be tensed or relaxed, and fingers get ready to ball into fists or stretch out into soft instruments of caress. Actions may become slow and ponderous, or sprightly and vivacious. Sensory muscles are affected: eyes move in their sockets, pupils dilate or contract, nostrils twitch (and, if we were cats, our ears would swivel and whiskers twitch as well). Eye contact is penetrating or steely, or the gaze averted or shy. And the brain sets up patterns of expectation and prediction: some constellations of attention, memory, thought and imagination become primed; others may fade into the background.

Many of these bodily and behavioural aspects of emotion are visible or audible, so emotions often have a communicative function, as well as being about personal well-being. Some of these signals have evolved biologically – as in ritualised displays of aggression, for example – while many others acquire significance in particular cultures. Hanging your head and avoiding eye contact signals subservience in some cultures, but insolence in others, so there is plenty of room for confusion, especially in a multicultural world. As a result of social learning, each of our emotions can become extreme, blocked or pathological, and thus generate additional problems.

Finally, emotions get us ready to resolve imbalances and unwanted conditions in our world: we see off a competitor, secure our 'heart's desire', or recover from a loss, for example. When the action is successful, the situation is rectified, and a sense of well-being is re-established. We are, we might say, happy. Happiness comes in many forms, but often it is a temporary state that follows from having put the world to rights. Because there are different kinds of threat and achievement, there are several kinds of happiness. Physically, happiness usually signals itself through a softening of the face and eyes, a smile or a laugh, and relaxation of the muscles of the neck, shoulders, arms and legs. I will try to illustrate each of these features of emotion in the next section.

Basic emotional modes

I should say that no one has a definitive list of what these emotional primitives are, or even how many. Estimates range from four up to a dozen. For the purposes of illustration, I'm going to offer you eleven, but other authors lump some of these together. There even remains some disagreement about whether such a list is possible at all. Nevertheless, a fairly broad consensus has emerged out of the work of American psychologists Paul Ekman and Jaak Panksepp, in particular.[4]

Distress mode

Young animals of many species, like human babies, have a built-in response to adverse physical conditions, strange and unexpected events, and abandonment: they cry out and they squirm in full-bodied attempts to summon help and find relief. Energy floods their muscles, lungs suck in big gulps of air and then pump them out through suitable configured vocal cords, and the on-board siren sounds. Eyes dilate and swivel so that any signs of rescue can be detected early. *In extremis*, this mode can tip into a strong apprehension of the imminence of death or serious injury, accompanied by exaggerated panic and a full-blown emotional and physical meltdown from which it can take a while for the system to recover. Hysteria and panic attacks are debilitating and distressing in their own right. As adults, what were the full-on distress signals of the baby become, in most cultures, substantially inhibited or disguised – but the consolation of being held by another person retains its power. Being held in a familiar firm but gentle way, by someone whose smell and sound are also familiar, signals the end of the emergency, or at least that help is at hand and things are being brought under control. This kind of happiness – the resolution of distress and relief at being rescued – we might call *comfort*.

Recovery mode

Here, I'll lump together the various systems that are triggered by the need to recover from exhaustion, illness or injury. Energy and activation are drawn away from the external senses and from the muscles involved in moving and manipulating, and are instead focused inwards on processes of repair and replenishment ('recharging the batteries'). During deep rest and sleep, wounds heal more rapidly and white blood cells are created. Depriving animals such as ourselves of sleep has increasingly dangerous effects on the immune system. During sleep, waste products are removed from the brain and other organs, and levels of ATP, the main energy-carrying molecule of the body, show rapid build-up in the brain.[5] For our ancestors, this passive, inward mode had its dangers, and the impulse to seek dark and private places to sleep and to 'lick one's wounds' may still be with us. If this mode becomes chronic or predominant it can turn against us and result in social isolation, or in debilitating conditions like chronic fatigue syndrome or ME (myalgic encephalomyelitis) where the body gets stuck in this recuperative state. The kind of happiness that accompanies the successful completion of recovery we might call zest or *joie de vivre*.

Disgust mode

This mode is triggered by the need to prevent toxic material entering the body, and/or the need to expel such material if it has already slipped past our guard. At a physical level, rotten food, dirt and body waste may trigger disgust, and in many societies this reaction generalises to things one finds morally or psychologically repulsive. In the expelling form of disgust, physiological reflexes take over and we 'throw up' the offending matter, or void it through the bowels. The facial expression of opening the mouth and sticking out the tongue (and, maybe, saying 'Yuk!' at the same time) symbolises this disgorging. The alternative gesture of disgust is the wrinkling of nostrils, mouth and eyes – sealing up the orifices to stop foul things entering. Behaviourally, disgust involves the tendency to withdraw; to distance oneself from a source of infection, for example. On the mental level, the brain might generate images of being contaminated or even poisoned. When disgust mode becomes pathological – chronic or excessive – it can turn into a generalised disdain for the ordinary activities of life, an obsession with cleanliness, or an attitude of misanthropy or contempt towards other people. The kind of happiness that comes with the passing of disgust involves a general reopening up to life and experience: a feeling of re-engagement, we could call it.

Fear mode

Fear is aroused by the appearance, real or anticipated, of something dangerous, accompanied by the possibility of avoidance or escape. In response to this class of triggers, the body mobilises for action. Breathing quickens and activating hormones such as adrenaline and cortisol, as well as energy sources, are released into the bloodstream. Large muscles, especially in the legs, are pumped up and ready to leap into action. (It may be that restless leg syndrome reflects an activation of this reflex, possibly in response to subliminal perceptions of imagined threat – an impulse to move and run in response to no obvious cause.) At the same time, blood is withdrawn from the digestive system, resulting sometimes in the fluttery feeling we call 'the collywobbles', or a sudden loss of the background 'hum' of activity in the stomach, which we may experience as a 'sinking' or 'hollow' feeling. In extreme cases there might be knocking of the knees, or loss of bladder or anal sphincter control, as more and more of the body's resources are conscripted to deal with the emergency (and less and less are left over to serve normal muscle tone). Attention is alert for avenues of escape or, failing that, places to hide. If there is nowhere to run or hide, strategies of appeasement or displays of powerlessness – showing one's belly in the case of some animals, for example – may be deployed. When fear becomes predominant, phobias and apprehensions of various kinds may become established. The kind of happiness that emerges when fear has passed might be called relief.

Anger mode

This mode kicks in when the body-brain judges that some of its assets and resources are under threat from another being who might be susceptible to intimidation. You hear footsteps downstairs in the middle of the night. A rival academic attacks your pet theory in a public meeting ... There is threat, in other words, but to run away would be to lose, in this situation, so anger mode intends to make the competitor back off or back down. Again, heart and lungs crank up to provide more energy, testosterone floods the system, and facial expression and posture signal aggression and the readiness to attack. Facial muscles tighten and the brow knots, giving us a more angular and forbidding expression. Hands ball into fists, shoulders tense, legs and pelvis brace. At a social level, these signals may recruit your own kith and kin, who rally round in support. (A playground fight quickly gathers a partisan crowd.) Chronic or excessive anger mode creates costs in the form of raised blood pressure and damaging amounts of ambient cortisol which suppress the immune system, as well as short-temperedness and a consequent loss of the goodwill of colleagues and friends. The successful discharge of anger results in the kind of happiness we could call triumph.

Sorrow mode

This mode is triggered by a different kind of threat: the loss of a stable, valued part of your world that is deemed irrecoverable. The loss of one's life savings in a pensions scam or the death of a partner may trigger the physiological and behavioural reactions typical of mourning. The loss could also be of a skill or capacity, such as the ability to work, drive, flirt, or live up to images of 'manliness', 'femininity' or 'self-reliance' as a result of accident or age. As with recovery mode (of which sadness may be an outgrowth), the pull is towards withdrawal, inwardness and a process of adjustment and reconceptualisation – 'coming to terms with' life without this familiar adjunct. Mourning the loss of a dropped ice cream, or of a holiday friendship, may lead to an unusually quiet child on the car ride home. The work of mourning the loss of a spouse may never be completed. Excess of sorrow becomes depression, of course. Relief from sorrow, like relief from exhaustion, may bloom as the kind of happiness that we can call 'coming back to life', or 'rediscovering one's zest for life'.

Shame mode

Human beings, like dogs and chimpanzees, are social animals, and being accepted by the pack or the troop has a direct effect on one's chances of flourishing, and even sometimes surviving. So it is small wonder that social animals experience the kind of threat to their social standing which we call shame. (Shyness and guilt can be treated as versions of shame. Shyness stems from the anxiety that one's appearance or social performance is not up to the job of securing one's social position. Guilt is the kind of shame that involves transgressing social norms and standards and thus risking being demoted or cast out.) Like mourning and recovery, shame is an introverted reaction, distinguished by the performance of social behaviour designed to appease or apologise. A hangdog expression betrays our similarity in this regard to 'man's best friend'. The head is lowered and eye contact withdrawn, as if one were acknowledging one's unworthiness to be connected – on a par – with others.[6] Behaviourally, shame leads to a loss of initiative and a reliance on others to indicate the next, or the correct, course of action. (What do I have to do to atone: to recover at-one-ness with the pack?). Chronic shame becomes self-consciousness or insecurity, and may result clinically in Social or Generalised Anxiety Disorders. The happiness associated with absolution from shame might be called acceptance or 'the relief of forgiveness'.

Desire mode

Not all the basic emotional systems deal with threats. I want to mention two that are about gaining benefits rather than avoiding losses. The first is desire mode, in which I will include sex and ambition as well as hunger, thirst, and the need to warm up or cool down. (Panksepp calls these the 'seeking' systems.)[7] As needs arise within the body-brain, so attention is primed to notice relevant resources and opportunities for assuaging these desires, and actions are instigated to create or reveal such desirables. Possible mates and prey – or internet dating sites, or suitably stocked delicatessens – become foregrounded in perception or physically sought, and, when sighted, attempts are made to secure the desired benefit. Relevant memories and experiences are mobilised, perfume applied, credit card limits checked – and the chase is on! The adoption of or obsession with insatiable desires leads, as we are taught, to several of the Seven Deadly Sins; while consummation of the desire leads to the (usually more short-lived than we expected) kind of happiness called satisfaction or satiation.

Enquiry mode

Another mode of approach, rather than avoidance, aims at the acquisition and consumption not of 'goods' but of knowledge and experience. The triggers for this are (a) internal activation of interests and ongoing problems/predicaments, and/or (b) the presence of an opportunity to explore a pertinent or novel aspect of the current situation. (Human beings are bristling with predictions and expectations just begging to be tested and investigated . . .) If a preliminary appraisal (by the body-brain) judges that the situation contains significant threats, fear or anger modes – flight or fight – might override enquiry (as they often do in classrooms, for example). But if things look interesting, safe and relevant enough, then enquiry mode leads you to make an (appropriately cautious) approach and initiate investigation. You prod it, smile at it, or open it and read it. An excess of enquiry mode can lead to chronic insecurity, in which too much of one's environment is seen as ambiguous or potentially treacherous. (Gregory Bateson suggested that such an over-problematisation of one's world is a feature of schizophrenia, for example.)[8] The happiness that follows learning we might call mastery or a satisfying sense of comprehending that which was previously obscure (and therefore potentially unpredictable).

Actually, enquiry mode and (sometimes) desire mode have a feeling of pleasure that accompanied them at the time, as well as a 'rebound' feeling of happiness on resolution. Mihaly Czikszentmihalyi has argued that 'the thrill of the chase', whether for material, mate or mastery, is itself inherently happiness-making. The state of being engrossed in the pursuit or the endeavour, which he calls 'flow', has evolved to be intrinsically motivating. If nothing else is calling on our time and energy, we are built to be both acquisitive and inquisitive.[9] Whereas the 'negative' modes get us to fixate on what is wrong, and how to fix it, in enquiry mode we are more open-minded and creative. We are built to explore and learn, to enhance our ability to 'know our way about', because you never know when such informed competence might be useful. On the basis of a good deal of research, Barbara Frederickson, author of the 'broaden and build' theory of positive emotions, says: 'positive emotions broaden people's modes of thinking and action, which, over time, builds their enduring personal and social resources'.[10] If you are not tired, afraid, angry, nauseous or sad, then do a crossword puzzle, go for a walk or see what's on television: you never know what you might learn and when it might come in handy. Soaps and reality shows feed this desire for vicarious learning – though whether repeated demonstrations of the costs of being untrustworthy actually feed through into changes of behaviour is, as far as I am aware, unproven.

Care mode

This mode is the partner to the distress mode. It is the response called out in us by hearing or seeing distress in others. Compassion is innate to human beings: small children show signs of sympathetic distress to other children, and are sensitive to adults' needs for help or reassurance. The hormones oxytocin and vasopressin are strongly implicated in caring mode, their release into the body being triggered by hearing a distress call or by breastfeeding, for example. Without our thinking about it, the maternal care mode results in softening and welcoming gestures, and in speech becoming slower, deeper and more rhythmic. A successful rescue results in a kind of happy contentment for both rescuer (e.g. mother) and rescued (child); other rewarding chemicals ('endorphins') are released in the brains of both.[11] When care mode becomes compulsive or excessive, parents or carers may suffer from 'compassion fatigue' or 'burnout'. They may also damage the development of the cared-for through over-protecting and prematurely rescuing them, so that the child, for example, is deprived of valuable opportunities to learn how to 'stand on her own feet', or to develop strategies for self-soothing. As with the other emotional modes, it is balance and appropriateness that are key.

Anxiety mode

Finally in this list of illustrations, there is one more mode we should note: anxiety mode. This is different from fear. In fear mode, the threat and the potential remedy are clear. You may be taking a moment or two to check your reading of the situation, but the inclination to escape or avoid are already activated. In anxiety mode – in the way I am using the term here – you are unsure which of the emotional modes it is appropriate to engage. Is it a trick or a treat? Can I escape (fear) or should I stay and fight it out (anger)? Is the loss really irrecoverable (sorrow) or are there things to investigate before I come to that conclusion (enquiry)? Can I find someone to blame (anger) or is it my fault (shame)? In these terms, anxiety – like all the other more focused modes – is often appropriate and intelligent. As American business guru Tom Peters once pointed out to an audience of executives, working in highly complex environments, 'If you're not confused, you're not paying attention'.[12] Indeed, like a pack of hounds casting around for the scent, we often cycle through several of the modes before we settle on the one that is the most promising base from which to attempt to pursue, or re-establish, our core sense of well-being. (Classically, coming to terms with bereavement, for example, involves cycling between denial, anger, shame and sorrow, before, if all goes well, adding to the mix a kind of acceptance and a resumed interest (enquiry) in engaging with life under the new conditions.)[13]

Note that I have said nothing yet about consciousness. All of these modes can operate perfectly well at the physiological and behavioural levels without any need for conscious supervision, or even awareness. People sometimes talk as if emotions were 'useful to us', enabling us to read our own inner state and thus regulate it better – like an airline pilot monitoring the readings on the bank of dials in front of her, and making rational decisions about what parameters she needs to adjust. Most of the time, however, the electronic signals that give rise to these 'readings' can perfectly well be fed back into automated control systems which compute the necessary adjustments without intervention by the human pilot – and the somatic autopilot does the same.

The bodily processes underlying some of these modes have already been well researched. It seems as if the different modes do not rely on entirely separate chemical and neural circuitry, so we should not expect to find a clearly defined 'anger centre' or 'sadness network' in the body-brain. That's why I refer to them as 'modes' and not as 'systems' or 'networks'. The emotions are a set of evolutionarily developed collections of processes into which the complex body-brain-mind interactions are inclined to settle (in the face of different kinds of event). Each mode draws, at least in part, on a common set of resources, but links them together, in the heat of the moment, into a signature constellation of bodily states and activities.

Out of this dynamic dance emerge two things: a view of the 'external world' and a summing-up of the state of the interior – 'How I'm feeling right now'. These two 'views' arise mutually. My sense of what the world is like shapes how I feel, and how I feel colours my view of the world. Hence the common conflation of the two in language. It's not that 'I am irritable'; you really are irritating. It's not that 'I am feeling queasy'; this burger really is disgusting. Emotions are projected into the objects around us. Conversely, the emotion we feel is powerfully influenced by the psychological spin we put on the context in which we find ourselves. People who have become anxious as a result of crossing a rickety bridge feel stronger physical attraction to a good-looking young researcher who stops them for an interview. They experience their arousal not as anxiety, linked to the bridge, but as desire, linked to the person they meet. (If they haven't just walked over the bridge, people don't find the same interviewer so attractive.)[14]

Earlier attempts to try to associate different bits of the brain with specific emotions do not hold up under scientific scrutiny. The amygdala, for example, has often been thought of as the neural centre for *fear*. However, it is now more accurate to see it as an area that is activated when the body-brain is having trouble deciding on what kind of a predicament it is currently faced with. The amygdala is more associated with my higher-level *anxiety* mode than with fear per se. Similarly, the insula used to be specifically associated with *disgust*, but (as we saw in the previous chapter) it actually has a more general role in integrating the views of the internal and external worlds. This is obviously relevant to determining whether 'disgust mode' needs engaging, but also relates to the wider process of deciding which out of all the modes is the most appropriate. Although the orbito-frontal cortex has been linked to the control of *anger*, it is more accurate to see it as having a wider role in the reorganisation of behaviour in the face of surprise, disappointment or frustration. And the old idea that the anterior cingulate plays a particular role in sorrow has now been superseded by the view that it has a more general part to play in deciding whether to respond to external events; and, if so, how. The decision, in the face of loss, to withdraw and become less responsive to the flux of passing events, and to focus instead on inner processes of adjustment and realignment, is just one example of this function.[15]

The complexification of emotion

These basic emotional modes form a set of primary emotional colours. Out of these, different cultures, different families and different individuals mix their customised palettes of shades and hues. The basic colours get overlaid and blended in a host of ways, and the accepted blends and displays change over time. In Japan, for example, anger is seen as highly inappropriate between relatives or colleagues, and legitimate only in conflicts between different social groups (as in the Samurai wars of feudal times). By contrast, displays of anger and assertiveness amongst people who know and like each other are common in many other cultures, such as the USA, UK and many Mediterranean countries. On the tiny Pacific atoll of Ifaluk, happiness is parsed very differently from the Western way. A low-intensity smile of contentment is OK, but any display of exuberant glee is seen as an egotistical – and therefore shameful – breach of social etiquette and you get a ticking-off. Signs of fearfulness are often treated as indicative of weakness in 'macho' Western cultures such as the army, or even in many schools; and the differences between the genders are less marked in this respect than they used to be. A maidenly reticence and hesitancy used to be seen as 'becoming' in girls and young women; much less so now.[16]

As with real colours, the primary emotions get blended into a glittering array of different moods and feelings. *Embarrassment* is a mild version of shame or guilt. *Jealousy* is a blend of distress and fear, caused by an anticipated loss, and aggressive intimidation designed to prevent the loss occurring (though it often has the opposite effect). Similarly *sulking* is simultaneously a hurt withdrawal and licking of wounds, and a (passive) aggressive attempt to punish the offender (by the withdrawal of affection or responsiveness) so they won't do it (whatever 'it' is) again. *Contempt* or *disdain* are social emotions that blend anger and disgust: one is repulsed by the weakness or inappropriateness of someone's social behaviour, and irritated with them for not knowing better. *Indignation* is a form of intimidation aimed at forcing someone, through a mixture of reasoning and bombast, to admit that their estimation of you (or of another being or belief you care about) is unjustified. *Nostalgia* is a mild form of sorrow, focused on the loss of what are seen as 'the good old days'. *Admiration* is the experience of desiring to possess some of the character traits – beauty, wit, kindness – of the admired person. *Envy* is admiration with a dose of disdain mixed in. And so on. We might note that, even though some of these blended emotions are commonly seen as quite 'refined' – hallmarks of cultured human beings – they can still be traced back to their roots in the body and the brain. Damasio and his colleagues have shown that admiration, as well as the feeling of compassion, activates brain–body loops that are associated with the regulation of the internal milieu (gut function, heart rate and so on). As our emotional lives become more subtle and sophisticated, they retain their rootedness in the physical workings of the body.[17]

Families encourage or prohibit different colours, and thus influence the palette which each child will take forward into life. My mother, for example, lived her life on the edge of panic. She felt that she was barely coping – certainly when it came to managing a small boy and a semi-wild Airedale terrier, in the midst of the London bombings of the Second World War. We lived in fear of tipping her over into one of 'Mum's turns': a hysterical meltdown that would take days to pass. Mum was doing her best, and it was our job not to question or object to it, whatever it was, so as not to trigger the underlying insecurity. This meant that any display of assertion or dissent by me was taboo; my job was to be Sunny Jim and radiate, as my mother put it, 'sweetness and light'. I am still in recovery from this inescapable training, finding it hard, to this day, to be assertive without becoming aggressive, especially with the people who matter to me most.

Alternatively, children's emotions can be tuned by their observation of adults. Emotional reactiveness is contagious. If Mummy makes a face of disgust when I start to investigate the food in the dog's bowl, I learn to attach my own disgust to similar sights and smells. The same happens between adults. I can shrug off having made a tactless remark if other people round the dinner table treat it as a non-event, whereas I will have a very different 'learning experience' if it is greeted with a stony silence, an icy look or a sharp intake of breath.[18]

Through these kinds of learning experiences, emotions can become distorted, convoluted and entangled. We may learn to be afraid of things we value, such as success or intimacy. I might have learned to treat my innocent delight at a success as shameful and somehow disrespectful of others (if I had been raised an Ifaluk). I might have learned to see imaginary slights in innocent conversations and thus startle my friends with sudden – to them inexplicable – outbursts of hostility or sulking. I could have come to find delight and excitement in activities (fetishes, for example) that some other people find disgusting. Anger doesn't become 'bad' or 'wrong' per se, but it can be deployed unskilfully. As Aristotle said: 'Anybody can become angry – that is easy; but to be angry with the right person, to the right degree, at the right time, for the right purpose, and in the right way – that is not within everybody's power and is not easy.'[19]

Emotional control and inhibition

As well as informal tuition in which emotions are acceptable and which not, children are taught to moderate the intensity and length of their emotional displays. Sometimes they are directly told 'That's enough now' or 'It's not that bad'. More harshly, they may be told that 'Big boys don't cry', or to 'Suck it up' and 'Get over it'. The famous 'marshmallow test' – if you can resist eating one sweet for ten minutes you are rewarded with a second – shows how useful these strategies for emotional self-regulation can be. A variety of measures show that people who were able, as children, to restrain the desire system do better in adult life.[20] However, as we shall see in a moment, the ability to control emotion can also have negative consequences, especially if strategies have congealed into unconscious habits that are no longer appropriate or helpful. What was once a short-term refuge from emotion can turn into a lasting kind of imprisonment.

There are a number of ways of moderating the activity of the different emotional modes. You can distract yourself by, for example, covering your eyes, singing a song in your head, or trying to interest yourself in reading a book or playing with a toy. But many techniques for emotional self-management directly involve the body. After all, e-motion is strongly connected with motion. Emotion's job is to prepare us for action; and emotions themselves often involve overt physical movement. Thus, as well as being capable of distracting ourselves, or damping awareness directly through cognitive control, we can also reduce emotional vibrations by clamping the bits of the body involved. Muscles in the abdomen can be stiffened to prevent the physical tremors that give rise to those butterflies. If the breath flutters and sighs with sadness, then other muscles in the torso can be tightened up to make breathing shallow and controlled. When people are feeling guilty or aroused by the effort to lie, their skin conductance goes up, but their overall body movements go down. Gaze and facial expression tend to become fixed and blank, for example.[21]

But emotional control via physical tension is a mixed blessing. If short-term, strategic and appropriate to the real situation, the ability to check one's emotional impulses and displays is a boon. Social manners and mores require it continually. But where the restraint of attention, or of bodily movement, becomes habitual, there are increasing costs. First, inhibition is energetically expensive. It is tiring – and stressful – being tense or phoney all the time. And this expenditure of energy has knock-on effects on other aspects of bodily functioning. The immune system can become less effective, leaving us less able to fight off infections. When the muscles of the diaphragm and abdomen are chronically tensed, the healthy working of the digestive system may be compromised and various gastrointestinal disorders ensue. The field of psychoneuroimmunology (PNI) is now well established, and the idea of 'psychosomatic' disorders as being somehow 'made up' is no longer credible. A wide range of illnesses have been reliably linked to emotional suppression, including asthma, heart disease, ulcerative colitis, and, not surprisingly, a variety of aches and pains such as back pain and tension headache. (More on this in Chapter 11.)[22]

Second, the inhibition of expressiveness – resulting in a tense or wooden demeanour – often has social consequences. Giving less of ourselves away and presenting a rigid persona, we become harder to know, and intimacy and warm friendship may suffer. We know that when we are in formal situations our body language becomes inhibited; but when we have become 'inhibited', in the everyday sense of the word, we carry that formality around with us.[23] The famous 'stiff upper lip' of the British army officer marching into battle will have had the desired effect of concealing the physical trembling that is associated with strong fear; but if the lip stayed stiff when he came home from the war to his family, he would have been harder to love (witness the moving portrayal of a war veteran by Colin Firth in the film *The Railway Man*). And many a quiet, gentle person has discovered in psychotherapy that behind their fixed smile and low, deliberate voice, there is a stagnant well of unfelt and unexpressed rage or resentment. In one telling piece of research, Martina Ardizzi and her colleagues studied the long-term effects of emotional repression in people who had lived through the traumatic conflict in war-torn Sierra Leone. Specifically, they worked with a group of so-called 'street-boys': adolescent boys who had lost their families in the conflict and were struggling to survive in feral gangs. When shown pictures of faces displaying different emotions, these boys' own faces showed less reaction – they had learned to inhibit their responsiveness to other people's emotion – and they found it harder to read the facial expressions correctly than a control group.[24]

Previous research has shown that people's empathy, especially their ability to read others' emotional facial expressions correctly, depends on their ability to mimic, at a very subtle level, the expressions they are seeing. If you record the activity in people's facial muscles as they are looking at photos of emotional faces, those whose own muscles resonate slightly with the emotion they are seeing identify it better. If you have learned to keep a poker face, whatever the stress of the situation you are in, you inhibit that resonance, and so are less well attuned to the state of those around you. The partners of habitual suppressors experience less rapport during their conversations.

This effect has even been shown in people who have had Botox injections to smooth out their wrinkles or make their lips fuller. The Botox stiffens your face, interferes with your own muscle function, and thus reduces your ability to show this sympathetic resonance. Botox mucks up the signalling from your face to the emotional loops in the brain and the rest of the body. Compared with a group who have had a different skin filler, Restylane, which doesn't affect the muscles, Botoxed people were less able to tell what the people in the photos were feeling.[25] And you don't have to suffer an injection in order to experience this loss of empathy. If you are asked to hold a pencil in your mouth using your lips, you have to pucker up to do it. It is almost impossible to do this and smile at the same time. Conversely, if you have to hold the pencil using only your teeth, most people stretch the corners of their mouth out in a way that looks rather like a smile, and so are less able to pout or look sad. Paula Niedenthal has reported that these two manoeuvres selectively affected people's ability to read sad or happy faces. If your mouth is set in a pout, you are better at identifying the sad faces and worse at the happy ones; if you are forced to smile, the reverse.[26]

The third potential cost of emotional inhibition concerns mental functions that we would normally consider 'intelligent': thinking, memory and problem-solving. If exerting control over our attention and our muscles, in the ways described, drains the body's energetic batteries, we might wonder if this will hamper other processes that also need a good supply of juice. Indeed it does. If you are asked to suppress all emotional reactions while watching an upsetting film, your memory for information that was fed to you while watching is worse. If the film involves an argument, you recall the details of the argument less well. And, after watching the film, you are worse at solving anagram puzzles. In general, people who are habitual repressers of their emotions also report more memory problems in everyday life. And if you think a lot about controlling your emotional behaviour, your memory is worse still.[27]

Ever wondered why you are drawn to that chocolate bar after a hard day – and why it is so difficult to resist? Roy Baumeister and colleagues have shown that there is a direct link between blood sugar levels and the ability to exert control over feelings, thoughts and actions. It takes a lot of energy to keep the prefrontal cortex pumping out all that inhibitory activation, so if you do not take more sugar on board, or find ways of releasing stored glucose into your bloodstream, you are going to run out of gas. Tasks that require focused concentration, holding a lot of different factors in mind, or resisting competing temptations, will suffer. It turns out that people who suffer from hypoglycaemia – low blood sugar – have more trouble with self-control. Juvenile delinquents and adults in prison are more likely than average to be hypoglycaemic. When ordinary people are running out of willpower, a few swigs of a sugary drink improves their performance. (There is recent research that suggests that you don't actually have to ingest the sugar; just a gargle with some sugary lemonade will do the trick. It may be that the lack of concentration reflects not just a general depletion of energy but a reallocation of resources by the brain to different tasks. The mere taste of the sugar in the mouth can be sufficient to get the resource-allocation processes to switch back to the task in hand. Either way, a change in the body definitely causes a rapid change to the workings of the mind.)[28]

Crying: a case in point

It is hard to understand why human beings cry. Many other animals have, as we do, lachrymal glands that lubricate the eyes. Tears wash out specks of dirt; they protect the delicate skin of the eye from damage, infection or the harmful fumes released by a cut onion; and the lubrication, unless excessive, makes vision clearer. But no other animal, as far as we know, is moved to tears by a Mozart aria, or by witnessing undeserved kindness between people we do not know. The ability to be touched and moved by such things must surely count as amongst our more refined – some would even say spiritual – capacities; yet they manifest not as reasoned understanding, but as the overflowing of a weak saline disinfectant fluid from the sockets of the eyes. What further proof could we want that our highest sentiments are intimately connected with the workings of the body?

Perhaps the first thing to note is that emotional tears accompany several of the emotional modes. They are not exclusively connected with either distress or sorrow. When a baby yells for help, tears may flow down his cheeks; yet they could just be mechanical by-products of the huge internal pressure that such yelling requires. (The muscles round the eyes contract to protect them from this pressure, thus squeezing out the tears. If the tear ducts are empty, howling can be dry-eyed.) And as adults we can bellow in pain without weeping. So we can dissociate crying from crying out. (As children learn to recognise their caregivers, their crying may become quieter if such a rescuer is in view.) Grief too may be accompanied by seemingly unstoppable floods of tears, but may also be dry. We can weep with frustration, with rage, with joy, with relief and with sympathy – as well as with sadness, pain and distress.[29]

It is as if the crying is a literal counterpart to an emotional overflowing; an inability to contain the intensity. We know that in the build-up to crying there is increasingly intense activity in the sympathetic nervous system; but the parasympathetic brake may also be 'on', trying to restrain a public display of emotion that we may feel to be 'weak', 'unmanly' or just inappropriate. As the urge to cry builds, we often feel that battle for control: breathing becomes shallow or shuddery; the throat constricts (so we feel a 'lump' in it); the mouth tries to clamp itself so as not to tremble; we swallow continuously, as if trying literally to suck back in the feeling itself. Then, if the emotion proves too much, we have to let go and the dam breaks. And, for some reason, that release of emotional intensity recruits the lachrymal apparatus and causes a flood. (The lachrymal glands in the eye sockets can be activated directly from the lachrymal nucleus in the brain stem, as well as by higher centres such as our old friends the insula and the anterior cingulate). Emotional tears have been found to contain stress hormones, such as ACTH (adrenocorticotropic hormone), so it is possible that, as well as being a physical release of tension, crying is also a quick way of dumping the excess of such hormones that has accumulated during the build-up. (Normal tears drain away through internal channels into the nose and mouth but, like guttering in a downpour, this system may not drain fast enough to prevent an overflow.)[30]

When the overflow becomes public, tears can and do act as a social signal. Unless we have reason to be cynical, seeing someone else cry softens our heart. It is hard to carry on being caustic or righteous when the person we were fighting with starts to weep. Just as they do for mothers, tears can flip the rest of us from, for example, anger mode into care mode. A recent study showed that smelling a woman's tears, without knowing her or even being able to see her, caused a drop in testosterone and an increase in oxytocin in male volunteers.[31]

Emotion and cognition

What are often called 'higher mental processes' actually sit atop a whole lot of emotional and visceral goings-on. That is not a nuisance or a design fault; it is a deep part of our evolved nature as intelligent beings. To recap: at the core of our being there is a deep churning flux of bodily concerns, capabilities and activities. Via a host of different pathways, information about this turmoil loops up into the brain and makes contact with the opportunities and affordances for action that seem to be available in the current environment. Out of this meeting emerges, moment by moment, a provisional decision about which broad emotional mode is most appropriate. We are set to rest, or flee, or intimidate or join the rescue party. From this mode, the specifics of thought, action and perception then emerge. Memories are retrieved, plans are laid, words are chosen, attention is swivelled in one direction rather than another . . . and finally our first-choice action is implemented and we wait to see if it has the desired effect. All this fleshing out of the details of our response occurs within the broad emotional/motivational frame our body-brain has selected, and it is this final 'dotting of i's and crossing of t's' that psychologists generally call cognition. But it is the whole process that deserves the name of 'intelligence', not just the finishing touches.

So it comes as no surprise to discover just how much emotion influences cognition. New York University professor Jonathan Haidt likens the relationship to that between an elephant and its mahout, its rider. If they have a good working relationship, the cognitive mahout can nudge and steer the emotional elephant to the advantage of both. If they are at odds, the best-laid plans of the mahout can be overturned by the stubbornness and strength of the elephant. Even at the best of times, their joint possibilities are always dependent on the mood of the elephant. The mahout can do nothing by himself.[32] If we are feeling less metaphorical, we can translate the elephant and the mahout into what we previously called System One and System Two (see p. 100). But whichever you prefer, please remember that there are not really two different systems competing for control. There is just a single interwoven body-brain system that translates momentary constellations of Needs, Deeds and See'ds into 'the best thing to do next' – and which sometimes, when it matters, recruits inhibitory processes to slow things down, check impulses and explore a wider range of options.

Contrary to the Cartesian model, the effects of emotion on cognition can often be beneficial. For example, our emotional mood significantly affects our memory. Up to a point, emotional and physiological arousal enhances memory. Hormone receptors in the brain, sensitive to concentrations of noradrenaline and insulin, influence the nature and the strength of memory connections that are being laid down in the hippocampus. Smart 'cognitive enhancing' drugs such as modafinil and nicotine have demonstrable effects on attention and memory. And people who are more sensitive to their momentary visceral state recall emotional words and pictures better than those who are less sensitive. On the other hand the 'stress hormone' cortisol interferes with memory for facts and events. High-fat meals seem, for some reason, to selectively impair spatial memory.

More intriguing are a range of research findings from Joseph Forgas at the University of New South Wales in Australia, which show that mood affects memory, learning and thinking in surprising ways. Broadly, if what you need to learn or remember somehow matches your current emotional mode, you will process it better. You spend longer attending to it, connect it with a richer set of information and understandings already in your memory, and so store it in forms that are more readily retrievable. In one naturalistic study, Forgas found that people out shopping on sunny days notice and remember less of what is going on around them than people on rainy days, and the key variable is whether the weather has put them into a bad mood. Being in a 'bad mood' can sometimes actually benefit cognition.

So if you want to learn about possible dangers in your world, do it when you are already anxious. If you want to ready yourself to take assertive action in the future, first recall some memories that put you into an irritable frame of mind. When people are irritable or sad, they are also more critical of what they are learning, and more attentive to detail, both of which may be useful and appropriate. On the other hand, if you want your students to engage with what they are studying in a more creative and holistic way, show them a YouTube clip of *Fawlty Towers* to warm them up and put them into a good mood.

These mood-dependent switches in cognition can be very significant in cases such as the validity of eyewitness testimony. Happy people not only pay less attention to detail, they are also more prone to incorporate misleading information, suggested to them after the event, into their actual memory of the event. Basically, when we are happy we tend to notice and recall what we think is plausible rather than what actually happened: both perception and memory are strongly tinted by our own prior knowledge, opinions and expectations. Alarmingly, when people are happy they tend to rely more on ethnic or racial stereotypes when making judgements about other people. Just Photoshopping a turban on to an unfamiliar face makes happy people more prejudiced and, interestingly, people in a bad mood less prejudiced.

So if you vary your mood, you will learn different things from the same presentation. Imagine a series of school lessons in which you are learning about the Holocaust, first from a sad perspective, then from an angry one, and finally from a position of shame. All of these emotional lenses will highlight different facets of the information, direct attention in different ways, prime different emotional reactions and, overall, lead to a rich and complementary set of mental representations.[33]

If you were in a bad mood, you could criticise some of these research studies on the grounds of artificiality. They often sit people in a psych lab and try to induce different moods, quite out of context, by simply showing them a few pictures designed to make them mildly sad, angry or happy. In the real world, one's fluctuating moods are linked to real concerns and experiences. Would people show a different relationship between emotion and cognition if the set-up was more realistic? Canadian researchers Isabelle Blanchette and Serge Caparos tested not memory but reasoning, in contexts that were either of real significance to their participants, or neutral. For example, they asked war veterans, people who had been involved in the 2005 terrorist bombings in London, and survivors of childhood sexual abuse, to reason about information that either related to their own history or did not. Though emotion is often thought to skew reasoning ability – by, for example, making people respond to their feeling about a statement rather than its logical validity – in these cases people reasoned more accurately about the events in which they had been involved than more neutral events – even when the 'conclusion' of the argument was hard to take.[34]

The authors suggest an interesting interpretation: that reasoning is hard work, and it is only when you care enough about the subject that you are sufficiently motivated to make the effort to 'get it right'. When your emotions are engaged, you mind enough to think well. Here again we see that intelligence cannot be nailed down to one way of thinking or another; it resides in balance and appropriateness. Of course we are sometimes swayed and confuse what we would like to be true with what is logically valid. But equally, we sometimes make logical errors simply because we are not engaged enough with the problem. Emotional involvement can swing us either way.

*****

One final example of the relationship between emotion and intelligence comes from a series of well-known studies by Damasio and his colleagues which look at the way decision-making improves with the build-up of experience. In what's called the Iowa gambling task, people sit in front of four face-down packs of cards. On each card, they are told, is a sum of money, either plus or minus. They are given an initial stake, and they either add to it or lose, depending on the card they pick. They can pick the top card of any one of the four packs and turn it over to see whether they have won or lost on that round. At first, they haven't a clue, but as they have successive goes, they gradually build up knowledge about how the four packs are arranged. Unbeknownst to them, the packs are carefully rigged. Two of them give some large wins early on, but gradually these are offset by larger or more frequent losses. The other two packs have smaller wins, but over the long run the losses are smaller, so it gradually dawns on the players that these two packs are the better ones to choose from. As people play, a variety of measures of their choices are taken. They are asked if they know why they made the selection they did, and if they had any 'gut feeling'. They are also checked for their skin conductance (EDA) response as they are deciding on their choice, which, as we saw earlier, is a subtle but reliable sign of increased arousal. The development of each of these indicators can be charted over the course of a hundred or so goes.

The usual result is that people start to favour the 'good packs' after about ten goes or so, and, at the same time, their skin conductance measures show that they are beginning to react physiologically to the 'bad packs'. If their hand hovers over a bad pack you see a blip of arousal, indicating a negative physiological reaction. At this stage, though, players are unable to tell you why they are making the choices they do. More slowly, they begin to report 'feelings' or 'intuitions' about the packs – which are actually quite accurate – but they have low confidence in these feelings. Gradually their choices home in on the good packs, but it is not till they are already getting good scores that they begin to be able to explain why they are making the choices they do. So: behaviour begins to improve at the same time as the body begins to react differently to the four packs, but only later is there any conscious comprehension of what is going on. Their bodies – skin reactions and actual physical choices – are learning faster than their minds.

Damasio, a neurologist by training, has found that people with damage to a certain part of the prefrontal cortex (the ventromedial area) behave differently in this situation. Like the rest of us, they are eventually able to explain accurately the different compositions of the four packs. But this knowledge never makes contact with their actual choices. They keep on making bad choices as often as good ones. And, very significantly, Damasio found that they never develop the bodily signals that seem to be guiding the choices of the unimpaired players. To explain this, Damasio drew on the somatic marker theory. The brains of normal people tag the experiences with the packs with a visceral note of the feelings they have induced, and these feelings, reactivated in a new situation, help to influence their behaviour. Activity in the body helps to guide and constrain the more conscious kind of thinking we associate with the 'mind'. Those visceral memory traces act as an essential glue that binds together our actual behaviour – the choices we make – with our conscious thinking and understanding.

If, as a result of brain damage, this tagging isn't working, your choices are more at sea. Without the bit of the brain where the somatic markers feed into the decision-making process, competence and comprehension peel apart. Without feelings and intuitions, abstract intelligence loses touch with the place within us, deep in our bodies, where our concerns, needs and values are held. We can become 'clever-stupid', able to explain, reason and comprehend, but incapable of linking that understanding to the needs and pressures of everyday life. Far from being a kind of unwelcome interruption to intelligent action, emotions are the bridge that connects our cleverness to our embodied core values.

Follow-up studies have shown that performance on the Iowa task is reduced, not just for people with obvious brain damage, but for those of us whose somatic markers are weak, or who are relatively insensitive to our own internal states. It is only people who show the raised skin conductance who perform well on the task. The weaker your visceral response, the longer it takes you to pick up on what is going on.[35] Interestingly, self-awareness makes a difference too. If your skin and heart show good responses to the bad packs, but your awareness of those responses is weak, you also perform more poorly.[36]

But we need to keep a balance. Aristotle is right: our emotions are often triggered by inaccurate beliefs, or pent up by acquired habits of suppression or avoidance. Emotional displays can be socially inept or disruptive, and people do indeed get things out of proportion, and get themselves all worked up about events that, from the outside, don't seem to matter that much. And of course people's reasoning is usually less dispassionate or altruistic than it might purport to be. We are often confused or mistaken. Deciding whether 'this' is something to approach or avoid is always a judgement call, and your body-brain, though it does the best with the information and the archives to hand (including its somatic markers), can get you to flee from harmless things (like people who only want to love you) and approach dangerous ones (like your uncle's pit-bull terrier). Sometimes the lessons of our personal history are unreliable guides to the present. Sometimes we have to learn, slowly and painfully, to re-set the triggers of our belligerence or our apprehension.

But emotions themselves are part of our intelligent equipment. The fact that they can become perverted or unreliable speaks to their cultural overlays, not to emotion itself. The emotional systems, and all the bodily activity that underlies them, are intelligent and considerate allies, doing their best to help us make our way, not on-board anarchists, forever trying to disrupt the show. As Damasio's research shows, we cannot be smart without them.

THE EMBODIED MIND

MIND, n. a mysterious form of matter secreted by the brain. Its chief activity consists in the endeavour to ascertain its own nature, the futility of the attempt being due to the fact that it has nothing but itself to know itself with.

Ambrose Bierce

At the end of the previous chapter, we had already started to look at the way the body affects mental performance – especially the way we use the lessons of experience to guide new choices and decisions. In this chapter we will delve into these kinds of effects in more depth. I will show that what we have traditionally thought of as the hallmarks of truly human intelligence also rely absolutely on events and processes in the body.

Human intelligence is traditionally associated most strongly with activities that possess some or all of the following characteristics. They are articulate: that is, they involve the use of language and/or other kinds of symbols such as mathematical or scientific. Second, they are abstract: they involve concepts that are less closely or less obviously tied to concrete experience. Third, they are rational: they involve processes of reasoning or analysis whose validity can be checked by other people. Fourth, they are conscious and deliberate: that is, they seem to take place in the well-lit workroom of consciousness under the control of the internal thinker or critic. Fifth, they are likely to feel effortful, rather than smooth and automatic.

The key question is: how much can we attribute these more intellectual abilities to the workings of the body? The Cartesian answer was: not at all. The science of embodiment offers two alternatives to the Cartesian view: a radical one and a moderate one. The radical view denies that 'mind' (in the sense of 'the organ of intelligence') is anything other than body. If we understand body well enough, the need for frameworks and languages that appeal to disembodied mental processes disappears. (Philosophers call this view *eliminative materialism*.) The more moderate view – which is the one I shall draw on here – would agree with Descartes that 'mind' in the sense of consciousness is pretty mysterious, but that, however it arises, it is always the accompaniment of *a bodily state which is the real workhorse of intelligent activity*. Consciousness is not a 'place' where intelligent processes happen; it is a particularly odd effluvium of perfectly explicable, material processes in the body-brain.

The moderate view is perfectly happy to retain mind language as a convenient shorthand for talking about complex human processes and experiences. As an analogy, you could, if really pushed, try to unpack the knowledge of biology in terms of the language of subatomic physics. You could. But most of what is interesting about botany and zoology, or even genetics, would be impossibly cumbersome to talk about if all you were allowed to use were terms like quark and lepton and 'the weak nuclear force'. The same, says the moderate view, applies to body-talk and mind-talk. Some of what is interesting about us humans can be well talked about in body terms, while other aspects are handled more elegantly by mind language. Mind language is particularly useful when we are talking about complex relationships within and between people, and the so-called 'emergent properties' that these relationships give rise to. We can happily – and quite legitimately – juggle the two languages as we try to find ways of talking about these interesting aspects of ourselves, without obliging ourselves to believe that there really are two quite different kinds of stuff. Let's see how this perspective helps us answer our key question.

The mental distillery[1]

We saw earlier that the body-brain is a compulsive predictor. It is designed to use its experience to adjust its own behaviour, so that, if something happens that resembles something that has happened before, the system anticipates that similar consequences will follow. The system, in effect, asks: 'If X was one of my current needs, and Y was the state of the world at the time, and I did Z, did Z work?' Was there a beneficial effect on reducing my hunger, cooling me down, or getting my manuscript accepted? If there was, my body-brain will adjust the strength of the synapses involved so, next time X and Y co-occur, I'm more likely to do Z. If not, it will tinker with the links and the loops so that I'll try something different next time. If these antibodies were effective at neutralising this kind of virus, I'll make more of them, so next time I'll be better prepared. If you smiled back and held me softly when I smiled at you, I'll smile at you again next time.

To do this, the body-brain system has to create generalisations or abstractions. Imagine a whole lot of people trying to draw freehand circles (of approximately the same size) on transparent sheets. They would all be different. But stack them up on top of each other and look through them, and the overlaps reinforce each other and stand out, while the random variations disappear into a grey background. That's what the nerves and chemicals of the body-brain are doing all the time. So 'Timmy here now' is a single sketch (of next door's cat); but a whole lot of 'Timmy here now' impressions will blend into a more general circuit signifying 'Timmy'. Timmy isn't a thing; it's an abstraction that allows prediction. And 'Cat' is the distillation of a whole lot of Timmys and Tiddles, and 'Animal' is the distillation of the whole lot of cats, dogs, horses, tigers and so on that I have experienced or seen pictures of in books and on a screen. If I only ever see Timmy in my front garden, the distilled circuitry will correspond to 'Timmy in the garden' – and I will have trouble recognising him if he turns up at kindergarten. If I pick up from my mother a fear of cats, every time Timmy comes round, my registered pattern will be 'Timmy + Be Scared'. Be Scared will be, in Damasio's terms, Timmy's 'somatic marker'. The important point here is that even 'Timmy', solid and furry though he is, is an abstraction from experience. There is no abrupt hiatus between 'concrete things' and 'abstract concepts'.

Some of this developing circuitry represents generalised scripts or scenarios for dealing with recurring situations, like the 'getting up and going to work' script, or the 'how to behave and what to expect in a restaurant' scenario. The scripts and scenarios link together sets of motor habits, perceptual predictions and feelings of value or utility, across widely distributed areas of the brain, so that when one part of the circuit is triggered (by noticing an inviting-looking restaurant, for example) a whole set of appropriate responses are primed and ready to go. You expect to get a menu; food to appear within a certain time period; the waiter to be friendly but not to invite himself back to your place for the night; money to change hands at the end of the meal; and so on. (Any of these expectations could be violated, of course, but in general the script helps life to run smoothly.) The restaurant scenario overlaps with the 'going round to friends for dinner' scenario and draws on many of the same skills and sub-routines, but the 'package' is distinct. (There would be trouble if you sent your plate back, or tried to tip your hostess as you left. I have a friend who was quite discombobulated by once being asked to write something in a 'Comments' book after dinner at a work colleague's house!)

Some circuits distil out a perceptual core of objects that recur across different scenarios, and attach to them a variety of behavioural options. Next door's cat shows up at a variety of times and places, but 'Timmy' is defined centrally by his size, tabby fur, distinctive mew and tendency to bite you if you tickle his tummy, and on this core I hang a variety of action options. I can stroke him, chase him, tease him, talk to him or offer him some elderflower cordial (in a spirit of enquiry), depending on my mood. These types of core are often called *concepts*. If a concept has one habitual 'use', then whenever I come across an example of it, my circuitry will automatically prime the relevant motor programs. If I see a 'stapler', I get ready to staple; if I see a 'beach', I get ready to roll out my towel and test the temperature of the water.[2]

As I learn, my body-brain distils thousands of these abstractions, and we can imagine a number of strata emerging in the memory-laden circuitry of the body-brain. At the bottom are the millions of individual momentary impressions, each one a unique constellation of perception, action and concern. Call it the Impression Bank (or, as I think of it more informally, the

Compost Heap). Out of the Impression Bank, through some automatic processes of statistical aggregation, emerges a host of increasingly abstract abstractions which enable me to navigate ever more successfully through the world. Given a new scene, and a recurrent intention, I (hopefully) get better and better at making the right move – one that makes progress towards a looked-for benefit. If I'm a jazz saxophonist, I get better at improvising riffs that satisfyingly express my mood while staying true to the title and motif of the piece. If I am a trader in financial derivatives, I get better at drawing on all my experience to guide me towards the risk worth taking, and away from the risk driven by a rush of testosterone.[3]

We could imagine the concepts arranging themselves according to various dimensions of similarity into an expanding atlas of maps. Call these the Concept Maps, in which 'cats' and 'dogs' are closer together than 'pet mice' and 'dinosaurs', and 'animals' includes all of those, but not Mum and Dad or me. In neural terms, this map stretches forwards along the underside of the brain from the back to the front.

A complementary set of abstractions is a compendium of skilled motor programs, somewhat decoupled from the particular concepts with which they were originally associated, arranging themselves into a variety of Habit Maps. As a skill or habit is found to recur across different scenarios, that skilful core becomes increasingly disembedded from any particular setting. It develops a wider sphere of utility, we might say. Instead of being arranged according to similarities between object concepts, this map groups together skills that are alike in their actions: skating is closer to rollerblading than it is to plaiting hair or writing an essay. The habit maps develop along the top or dorsal pathways of the brain through the parietal lobe to the motor and premotor areas.

There is another kind of abstraction that distils out our visceral reactions to otherwise disparate events. We might call this set of visceral abstractions the Values Map, perhaps. Values Maps retain and highlight the emotional or somatic tone that the original experiences had in common. So we can develop concepts that separate out 'fair' from 'unfair', 'safe' from 'dangerous', 'kind' from 'unkind' or 'reliable' from 'untrustworthy'. There is research to suggest that these visceral echoes underpin what we usually consider to be our more abstract concepts. It is not that these concepts are nebulous or ethereal; it is just that they embody a feeling connection rather than a perceptual similarity or a common way of engaging. Out of experiences that involve me being treated *fairly* or *unfairly* begins a distillation process that will end up as the abstract concept we know as *Justice*. Out of a whole range of experiences that involve my being *rescued* and *comforted* emerges a core element of *Love*. Out of the sadness and anxiety that came from experiences of being *let down* or *betrayed* may emerge the underpinnings of abstract ideas like *Loyalty* or *Trust*.

Doubt, for example, is normally taken to be a rather abstract idea, implying a lack of belief or agreement. But it is also a vividly embodied experience of hesitation, arrested progress, bodily tension, frustration, and anxiety perhaps, which can be felt in the stomach, the pattern of breathing, a furrowing of brows, a pursing of lips or a quizzical cocking of the head. If someone asked you to mime doubt, I don't think you would find it hard.[4] And you would recognise that feeling when someone makes a good challenge to a cherished point of view, when you are not sure how to resolve a conflict with your partner, or when you come to an impasse in a work-in-progress. Mark Johnson suggests that we may become more or less insensitive to such 'cognitive emotions', but that 'once you start to pay attention to how you feel as you think, you will notice an entire submerged continent of feeling that supports, and is part of, your thoughts'.[5]

In support of this line of thought, a recent study has shown that some abstract concepts do indeed retain relatively more of their somatic savour than do more concrete ones. In an unusually well-controlled study, Gabriella Vigliocco, Stavroula-Thaleia Kousta and their colleagues found, contrary to expectations, that many abstract words were recognised faster than concrete words. We tend to think of abstract things as more difficult, but in this study they turned out not to be so. On closer inspection, they uncovered another surprise: that the reason for the quicker recognition of the abstract words was their greater emotionality. Of course, there are many different kinds of abstract concepts and some are indeed abstruse and bloodless. (I certainly don't have much emotional resonance with some of the abstract things I learned at school like *cosine* or *past participle* – though I do recall the lovely images of a rather fearsome *gerund* (reproduced as Figure 9 below) from Geoffrey Willans and Ronald Searle's 1954 book about school called *How to Be Topp*.)[6] But

Fig. 9 The gerund attacks some peaceful pronouns. From *How to Be Topp* by Geoffrey Willans, cartoon by Ronald Searle.

Vigliocco's list of abstract words included BARGAIN, BEAUTY, LUXURY, PROTEST and WELCOME, and you, like me, will probably have emotional reactions to all of these. Especially when we are learning new abstract words, their resonance with our own somatic concerns may provide the initial anchor-point in our minds, and this feeling-tone stays with it.[7]

Real events always involve a mix of perceptions, actions and concerns, but these maps, with their different kinds of abstractions – concepts, habits and values – enable us to partially decouple those different elements. We become able to look at objects with a more dispassionate eye. We can wander round an art gallery just looking, without needing to do anything about what we see. We develop the ability to ponder on our feelings and motivations in the absence of immediate calls to action (as in a counselling session or a heart-to-heart conversation). We can act 'for the joy of it' without having to evaluate the consequences of our actions (as in unselfconscious dancing). We can explore or rehearse possible courses of action 'off-line', and run simulations of how we would feel under different conditions (and thus be more considered and more creative). So these different distillates of experience enable a wide variety of 'higher' forms of intelligence. The (partial, variable) ability to decouple perceiving from desiring, desiring from acting, and acting from perceiving brings added freedom – and added complexity. If it seems to us that the world is full of more or less neutral 'objects' (and people), with which we can engage in various ways, it is because that is one common way in which the body-brain separates out its facets and ingredients. But this is an acquired ability, I suspect, not a fundamental design feature of the body-brain-mind.

*****

Abstractions vary in degree, as well as in kind. Obviously, some of these abstractions retain a good deal of the original sensory and motor detail, as well as the visceral markers of 'good' and 'bad'. 'Timmy' retains more of his concreteness than does 'animal'. Other abstractions, like those thrice-distilled vodkas which claim proudly that they are so pure they taste of absolutely nothing, lose much of their sensory and their motor flavours and retain just the underlying essence of the relationships or structures that were common to different events – the concepts *relationship, structure* and *event,* for example. But it seems likely, from the research we will look at in a moment, that even quite abstract abstractions retain their connections with their bodily roots, and these roots are easily – maybe even automatically – recoverable.

We might imagine two kinds of connection that each concept potentially has: lateral ones, that embed it in a network of meanings 'at its own level', so to speak; and vertical ones, that enable it to reactivate – or be reactivated by – the concrete actions, feelings, images and impressions from which it may once have been distilled. There might be a host of factors that determine whether, in any particular case, those somatic roots will actually be revived – but they could be.

Language

At this point we need to draw language into the story. Imagine another layer of connections that overlies and links to the planes and maps we have introduced already. Let's call it the Word-Scape. Words are ways of activating neural circuitry (and altering biochemical processes) through speech and writing. During the first year of their lives, children develop specialised circuitry for recognising and producing the sounds of the language in which they are immersed. Within the world of speech sounds, their powers of abstraction begin to pick out certain recurring sound shapes (called phonemes) and, guided by their parents, children learn to associate (groups of) these with objects and events that are being foregrounded for their attention. So the scenario associated with a toddler's nightly bath, for example, now has new circuits added to it. There is the sight of a familiar smiling face, the pleasant feel of the warm water on her skin, the need to squash up her face to stop suds getting into her eyes, the behavioural rituals of splashing and being snuggled in a towel, the anticipation of being put to bed afterwards ... and now there is another loop connecting that whole package to the sound pattern 'bath'. When a one-year-old hears that sound, the rest of the package is primed in pleasurable anticipation.

Neuroimaging studies have shown that words like CHAIR (or BATH) are represented in the maturing brain not in some neat cortical dictionary, but in terms of widespread loops of interconnected neurons that knit together (a) the muscle patterns you use to say the word, (b) the perceptual processes you use to recognise it, (c) what most chairs have in common, (d) what various kinds of chair look like, (e) what you typically use them for, (f) how CHAIR relates to THRONE and HAMMOCK, and (g) some memories of special chairs ('Granddad's favourite armchair') which you have personally encountered. Just seeing the word CHAIR primes the bits of the brain that control sitting down.

Having learned to map sounds on to articulatory movements, children can produce increasingly well-understood versions of these sound shapes for themselves, and thus discover the benefits of joining their linguistic community. Soon there will be another circuit, hooked into the BATH scenario, connected to a specific set of actions within the baby's own chest, lungs, throat and mouth, that can result in an approximate utterance of 'Bath!', and a lot of satisfyingly proud attention from the adults around. A little later, the Word-Scape will begin to cluster into 'thing-naming', 'action-naming' and 'want-naming' words, and new frames (the beginnings of grammar) will enable the child to experiment with creating novel combinations of a Thing name, Action name and Want name – and see what happens. 'More bath!' 'Katy bath!' 'No Katy – Daddy!!' And so on.

To get a foothold in the child's body-brain, words have to point to concepts, actions, feelings and scenarios that are already there. The game of using words to name, point and get others to follow orders is rapidly understood, and the child soon discovers that she can be both the agent and the recipient of these suggestions and injunctions. With the development of syntax, and an expanding vocabulary, however, she can be told about things and events that she has not yet experienced. Other people, through the medium of language, can start to shape and train the concepts to which her words originally applied. Words enable you to joint and assemble experiences in finer ways: not just a horse but a *green* horse; not just a lorry but a *slow* lorry. And so you can recombine the named features in new ways. Katy has seen, and named, her toys – a green horse and a pink tiger – and can now generate (and understand) a story about a pink horse. She has never seen a unicorn, but unicorns (and dragons and wizards) become conceivable.

So the Word-Scape begins to take on a life of its own, gradually detaching itself from its exclusive reliance on the experience-based abstractions formed automatically by the body-brain. It can incorporate distinctions made by the culture but not clearly marked in her own experience. At school, she will be told that Mum and Dad *are* animals and so is she, and she will protest that we aren't, we're *people*. As William James summed it up, in his own gloriously idiosyncratic language: 'Out of this aboriginal sensible muchness, attention carves out objects, which conception then names and identifies for ever – in the sky "constellations", on the earth "beach", "sea", "cliff", "bushes", "grass" '. Out of time we cut "days" and "nights", "summers" and "winters".[8] (I love the idea of the 'aboriginal sensible muchness', and I am glad we have plenty of it in England and don't have to go to Australia for it.)

That's a highly condensed overview of how abstract and linguistic concepts arise in the embodied body-brain. Obviously there aren't separable 'planes' or 'maps' to be seen inside the skull. Distinctions are not so clear-cut, and they are all represented functionally, in the way electrical and chemical activity gets shunted around, not (or not much) in terms of structural architecture. But the sketch will give us something on which to hang the ensuing discussions of how embodied abstract cognition can be possible. The body is not only intelligent in its own right; it makes possible the psychological abilities to which the term 'intelligence' is more usually applied.

Metaphor

Reason has grown out of the sensory and motor systems of the brain, and it still uses those systems, or structures developed from them ... The peculiar nature of our bodies shapes our very possibilities for conceptualisation.

George Lakoff and Mark Johnson[9]

So, one way in which abstract concepts might be functionally grounded in bodily concerns and experiences is through their historical (but still potentially active) association with clear somatic markers of various kinds. 'Injustice' may make your blood boil; 'vulnerability' may be a characteristic that makes you feel warm and caring towards someone; and you may, in the words of singer Alison Moyet, literally become 'weak in the presence of beauty'.

But there is another way in which this grounding can happen, and that is through metaphor. George Lakoff and Mark Johnson have shown just how much of our apparently abstract language is metaphorically derived from the basic properties and experiences of the human body, and of our concrete experience. We have bodies that *grow*: we used to be *small* but now we are grown *up* we are *bigger* and *taller*. We use hands to *grasp* things and feet to *kick*. We *fall* over and *skip* along. We feel *heat* and *cold, softness* and *roughness*, and *taste* sweet food or *smell* bad odours. We physically *travel* through space and go on *journeys*; we *leave* things and places behind and *arrive* at *destinations*. We get *in* and *out* of cars and lifts. Sometimes we are *close* to home and sometimes *far* away.

All of these concrete, embodied actions and experiences get metaphorically co-opted by language to help us think and talk about more abstract things. As we grow older, so we know that ambitions *grow*; that a little money is *small* change, *big* issues are important and *tall* stories are not to be believed. We *grasp* arguments, *kick* habits and now and then feel *down*. We *fall* in love and *skip* the boring bits. We get into *heated* discussions and give people the *cold* shoulder. Planes (and economies) make *soft* landings and we have *rough* nights. People are *sweet* to us, but sometimes their ideas *stink*. We go on *leave*, put our troubles *behind* us and are *destined* for better things. We get *into* hot water and are *out* of luck. We feel (emotionally) *close* to people, but are *far away* in our own thoughts. George Lakoff and Mark Johnson have *charted exhaustively* the *extent* to which all languages are *saturated* with what they *call* these '*dead* metaphors'. (OK, enough with the italics; you can add your own from here on.) We understand increasingly abstract ideas by raiding the vocabulary of the physical, and bending and combining literal meanings into new usages. And we do it all the time. We didn't use to be able to use a mouse to scroll, or surf the net. Trolls used to live under bridges; they're now more likely to be found in smelly bedrooms.

Historically, our abstract lexicon has grown over time by cannibalising more concrete images and experiences. Developmentally, children's entry into their linguistic communities depends on their recapitulating the same process of abstraction and distillation (though much speeded up). But it remains an open question how far these connections remain active. It could be that, as Wittgenstein once suggested, we kick away the ladder after we have used it to climb up to more elevated conceptual planes. (I can feel those italics wanting to creep back . . .) It is possible that 'justice' and 'gerund' need concrete experience to get going but can then cut free. Whether they 'cut the ties that bind' or stay connected is an empirical question . . .

Body talk

Arthur Glenberg, head of the Embodied Cognition Lab at Arizona State University, has carried out a number of experiments to explore how much the body is routinely involved in the way we understand language. In earlier studies, he showed that the action that you use to signal your understanding of a sentence interacts with the meaning of the sentence. Suppose you are shown on a screen a number of sentences, and you have to respond one way if they are sensible and another if they are scrambled or nonsense. If it makes sense, Bill has to push a lever away from him, let's say, while Ben, to indicate the same judgement, has to pull the lever towards him. What happens when the sentence is 'Anna *took* the pizza from Jill'? The action of 'taking' implies drawing something towards you. Without making any conscious connection between the direction of the action depicted in the sentence and the direction of their response, Ben responds to the sentence significantly faster than Bill. If the sentence was 'Anna *gave* the pizza to Jill', Bill would be faster than Ben. When the action you have to make is compatible with the meaning of the sentence, you speed up. When it conflicts, you are slowed down. Even though the task does not ask you to act out the sentence, or do anything that depends on its meaning (other than move the lever), the motor circuits of the brain cannot help it – they automatically prime themselves to carry out the action described. Simply reading or hearing a word primes its habitual use – even if this use is not referred to or even implied in the sentence. If you read 'Jane forgot the calculator', your brain automatically readies itself to make the kind of movements involved in pressing keys. If you read (as you are now doing), 'Mike spotted the kettle', your brain quite involuntarily activates the gesture of grasping a handle.[10]

The same correspondence between word and brain applies to sensory as well as motor language. If you know the smell of cinnamon, just hearing the word (or reading it, as you just did) is sufficient to activate olfactory areas of the brain. Reading the word 'telephone' automatically rings bells in the auditory processing region of the temporal lobe. And so on. That automatic priming, says Glenberg, is part of what understanding language involves. There is an involuntary tendency for the activation caused by hearing or seeing the word to shoot back down the vertical links we talked about earlier, and prime the relevant physical associations – and vice versa.

In a well-known study by John Bargh and colleagues, people who had being doing a task that involved unscrambling sentences in which some of the words related to stereotypes of the elderly (*grey, wrinkle, forgetful,* etc.) actually walked more slowly down the hallway after the 'experiment' was over![11] You can show the same relation the other way round as well. If you electrically stimulate the part of the brain that controls leg movements (with a technique called transcranial magnetic stimulation, TMS) you will more quickly recognise the work 'kick' than the word 'throw'. If you stimulate the hand regions, the difference is reversed. The Word-Scape, Concept Maps and Habit Maps seem compelled to share their activity with each other.

Well, you may not find that very surprising. After all, the sentences used in these two studies described concrete, physical movements, so it is plausible that there might have been some involuntary leakage between the language-processing bits of the brain and the corresponding motor-control bits. But suppose the sentences had been 'Anna *gave* the plot away to Jill', or 'Jill *took* the news badly'. Here, the verbs *give* and *take* are being used metaphorically; nothing solid actually changes hands and there is no reason for the actors in the sentences to move anything but their mouths. Yet here too Glenberg found the same effects. 'Giving away' a plot twist or a punchline does not involve a physical action, yet our brains automatically fire up the circuitry that is involved in moving the hands away from the centre of the body.[12]

You see the same firing-up of the sensorimotor cortex when the critical verb does not even have a literal counterpart, as in 'Anna *delegated* responsibility to Jill', or 'Molly *told* Patrick the story'. The brain representations of giving and taking are still activated. Glenberg suggests that, during childhood, we learn a variety of 'action schemas', such as giving and taking, that gradually become distilled by the brain into more abstract patterns such as 'Source – Object – Recipient – Mode of Transfer'. He says: 'Given that the action schema initially developed to control the arm and hand during literal, concrete transfer, when the action schema is contacted during the comprehension of abstract transfer sentences, areas of motor cortex controlling the hand also become active.'[13] Yet again, by the way, we see the motor cortex getting in on the act of perception. We make sense of the world, even when it is rather abstract, by getting ready to act on it or interact with it. The sensori-motor part of the brain provided the original platform for the development of more abstract cognition and comprehension, and continues to do so throughout life.

If you look at the structure of the brain, you can see that the areas that are traditionally associated with understanding and producing language are intricately tied up with concrete actions. One of these is called Broca's area, and neuroimaging studies have shown that, as well as its involvement in language production, Broca's area is active when concrete actions are being observed, or even when they are being imagined. When Broca's area is damaged, people lose the ability to construct grammatical sentences and their speech becomes cryptic and 'telegraphic'. They also become unable to sequence musical tunes. It looks as if the ability to construct the complex sequences of grammar may well have piggybacked on a pre-existing brain area that performed the same choreography for non-verbal actions – and still does.[14]

How the body affects thinking

Our bodies do not just intrude on the way we understand and produce language. They influence our attempts to interact intelligently with the world around us. The interconnection of the abstract and the concrete reveals itself in how we behave, as well as in what is going on in our minds. We can change the way we think, feel and behave just by changing our posture or our breathing, for example. I can quickly make myself feel sad or frightened just by deliberately making deep sighs, or by breathing in a shallow, juddery sort of way. (If you have never tried this, have a go. It can be quite dramatic.) Intriguingly, a study from Turkey has shown that one important aspect of people's intelligence as measured by a standard test (Cattell's Culture Fair Intelligence Test) is significantly correlated with the depth of their normal (unselfconscious) breathing.[15] Even more startling, another study has found that people make better choices when they are controlling the need to pee! The theory is that the exertion of self-control spills over (as it were) from the physical realm into the cognitive, and enables people to restrain their more impulsive sides.[16]

Telling people to adopt different postures has a rapid effect on their bodies and their minds. When people are induced to slump in their chair, they feel less pride in their performance; they think less carefully about problems; they take their own thoughts less seriously; and they show less resilience in the face of difficulty.[17] Conversely, sitting up straight and crossing your arms makes you feel more stubborn. People with crossed arms persevered twice as long in solving anagram puzzles and were more successful than people with their arms by their sides.

Dana Carney and her colleagues had people adopt either a 'high-power' (relaxed, upright) or a 'low-power' (slumped, constricted) pose for two minutes. Then they took part in a gambling game where they could play either safe or risky. Of those who had been standing or sitting in the high-power posture 86 per cent took the risky bet; only 60 per cent of the low-power people did so. Deliberately adopting the posture makes that association come true. Carney showed that the posturing – only two minutes, remember – also affected hormone levels. In the confident posture, testosterone increased by 20 per cent and cortisol (the stress hormone) went down by 10 per cent. In the submissive pose, the reverse was true: testosterone went down by 25 per cent and cortisol up by 15 per cent. Postures are associated with more abstract personality traits such as confidence or timidity. There is a non-arbitrary relationship between the idea of standing upright, with its connotations of being tall, proud and unashamed, and the metaphorical idea of an upright or upstanding character. And, by the way, the magnitude of these effects is different for different people. Some people seem to have managed to disconnect their minds from their bodies more than others.[18]

An intriguing example of the way so-called higher faculties are linked to bodily processes comes from another recent study by John Bargh's lab.[19] They wondered if there was any truth in the old saying 'Warm hands, warm heart'. Was there a real connection between physical warmth and the psychological state of feeling 'warmly' – i.e. friendly and trusting – towards another person? They had students do a product evaluation on either a warm or a cold therapeutic pad (the thing that athletes use for minor injuries), and then, for their help, they were given a small reward that they could choose either to keep for themselves or to give to a friend. The students who had handled the cold pad were three times more likely to keep the reward. Of those who had the warm pad, slightly more than half chose to pass the gift on to a friend.

The effect carried over into how warmly they rated another person's character. After having been asked to briefly hold a hot cup of coffee for the experimenter while he made some notes, students judged a hypothetical character as more likely to be 'warm and friendly' than if they had held an iced coffee. In neither experiment were any of the students aware of the correlation between the physical temperature and their judgement and decisions. This crossover is found in real life, too: people who say they feel lonely take more warm baths and showers. And it also occurs in the reverse direction. People who have experienced being socially rejected rate the temperature of the room they are in as colder than do people who haven't.

There is direct evidence that the physical and psychological judgements of warmth are processed by the same area of the brain – the insula again. In animals, the insula is clearly involved in processing both temperature and touch sensations. But in human beings, the same region has also been shown to underpin the sophisticated social emotions of trust and empathy on the one hand, and guilt and embarrassment on the other. The insula becomes more highly activated after experiences of social exclusion or rejection, for example. The structure that was designed to regulate bodily temperature now monitors the 'social temperature' as well. And it is not hard to imagine how and why this association might have come about. When we were babies, we (nearly) all experienced the physical warmth of our mother's body as closely associated with the crucial primordial feelings of being loved and secure. Being comforted and reassured usually goes along with the experience of warm bodily contact and stroking. In Harry Harlow's famous experiments in the 1950s, orphaned baby monkeys preferred to cling on to a dummy mother made of warm cloth than to the wire dummy that was actually the source of their milk. So we learn the metaphorical association between physical warmth and social warmth in our mother's arms. That association gets welded into our brains – and it stays with us for life.[20]

New studies are being published every week that demonstrate this cross-talk between body, behaviour, feeling and thinking in a host of ways. We use the word *burdensome* to refer to something that is *weighing heavily* on our mind, or which we are finding *hard to bear*. So it should not surprise us to learn that people carrying what they feel to be burdensome secrets (for example, having had an affair) are less likely to volunteer to help you move boxes of books; and they rate carrying a bag of shopping upstairs as requiring more effort. We talk about *washing our hands* of responsibility – and sure enough, in what has come to be called the Pontius Pilate Effect, physically washing your hands does actually reduce how guilty you feel about some misdemeanour. And if you have to copy out a passage of text that you find *distasteful*, you subsequently rate a strange-tasting drink as more *unpalatable*. To emphasise the reliability of what we are saying, we may use the expression *hand on heart* and, yes indeed, if we are told to place a hand on our heart, we really do behave more honestly. People who have just come up an escalator give more generously to charity. People touched briefly on the back by an interviewer report having a reduced fear of death.

Mathematics

Doing mathematics is, without question, one of the 'higher' forms of human thinking: one of the accomplishments most associated with the Cartesian view of intelligence. Mathematics is a world of abstract entities that make patterns. Yet even here the body and the physical world cannot be ignored. Arithmetic abstracts the property of number – numerosity – from groups of real things, sets up basic rules for manipulating numbers, and then explores how this system of ideas behaves. Geometry deals with mythical beasts – lines that have no thickness but can go on for ever; points that have no size; angles that are perfectly 'right' – and uncovers, within this abstracted world, what 'goes' and what doesn't. Algebra takes real objects and peels away everything about them except their very existence, and uncovers the lawful patterns you can make with xs and ys. Who would have thought that the area of a triangle was always half the base times the height – but it is. Who could possibly have imagined that, in Algebra Land, a single formula could tell you the value of x in any quadratic equation $ax^2 + bx + c = 0$? Remember this equation?

$$x = \frac{-b \pm \sqrt{b^2 - 4ac}}{2a}$$

Abstract though mathematics is, like language, it has its roots in the mud of everyday embodied experience. One such root is counting. There is a region in the parietal cortex which, if damaged, results in *acalculia*, an inability to manipulate numbers and perform arithmetical computations. This region is intimately connected with the nearby map of the fingers, and George Lakoff and Raphael Núñez have argued that the super-structure of mathematics is indeed underpinned by the child-hood foundation of counting on one's fingers. Set theory and the use of Venn diagrams rely on the everyday experience of putting things inside other things. If I put some marbles in a cup, and then put the cup in a saucepan, the marbles are, visibly and tangibly, all inside the saucepan.

If I have two friends, Jack and Jill, and I know from the evidence of my senses that Jack is taller than me, and Jill is shorter than me, then I don't need to be told that Jack is taller than Jill. All I need to be able to do, as children can at seven to eight years old, is see that I can replace 'me' with anybody – or with 'x' – and the same logic applies. By eight, I have become, without any training, an embodied logician. I have mastered what is called 'transitive inference'. Through stories, children get used to talking and thinking about things they have never seen. Tell a six-year-old a story about 'wuggles', and she will very happily tell you that, if three wuggles meet up with two wuggles, they make a group of five wuggles – without having a clue what a wuggle is. (Though, interestingly, if you ask her to tell you what 'three million' added to 'two million' is, she probably won't. Why? Because, she will tell you, 'We haven't done "millions" yet'. As they say, go figure.)[21]

When they meet arithmetic in school, children often get confused about the possibility of 'negative numbers', because their underlying embodied metaphor for adding and subtracting involves collections of things. Obviously you can't have less than 'no things', so '7 – 4' is perfectly OK – you are left with a smaller group – but '4 – 7' makes no sense at all because how can you have 'minus three things'? However, if your underlying metaphor for adding and subtracting is different – if it involves steps along a path – you will have a much easier time with negative numbers. If I walk four steps (from my starting point, or 'origin') to the right, and then seven steps to the left, I end up three steps to the left of where I started. That's perfectly sensible – and you can even call it 'minus 3' if you want and I won't mind.

A similar confusion underlies another common difficulty older students often encounter when they are being introduced to differential calculus. It involves how you (unconsciously, metaphorically) think about what a line is. One embodied idea is that a line is the trace left by a moving object – wheel tracks in soft ground, for example – or just a road. We say things like 'The highway *runs* from Oxford to Banbury' or 'The path *goes through* Mr McGregor's garden'. On this view, a 'point' is something additional to the line. It is 'on' the line, like a stone marker placed on the path. This is the intuitive view that most students have.

But there is another view, which sees the line as *made up of* points, like flagstones, or pixels perhaps. The line only looks continuous when the paviers are very small and close together. This is the view – technically called the Cauchy-Weierstrass definition of a line – that underpins the way most students are taught calculus. But they are not told that they will need to shift their 'embodied metaphor' for thinking about what a line is. They are told that the new view is more 'rigorous' or 'formal'; not just different from their intuitive view but somehow 'truer'. But because they continue to try to make sense of the idea of a *differential* in terms of the old metaphor, it doesn't make sense.

Actually, it is worse than that, because the whole idea of *space* changes. In Calculus Land, space isn't a continuous dynamic arena in which things happen (and leave traces); it is a static grid made up of numbers. Calculus turns space from geometry into arithmetic: '$y = x^2$' isn't a track in a field; it's a way of defining a very large set of pairs of numbers, x and y, that are linked by the rule. (I can still remember the confusion of trying to follow Mr Leonard's attempts to explain that there were two points P and Q 'vanishingly close together' on a curve. The atmosphere was lightened for a class of teenage boys only by his insistence – I can't now remember why – that 'this point Q has a certain Q-ness about it, while the other point P retains its characteristic . . .' You get the idea.) The point is: calculus is difficult precisely because it contradicts an unrecognised, but powerful, set of embodied assumptions; and this contradiction is not acknowledged and addressed because everyone is busy pretending that mathematics doesn't have such underpinnings.[22]

Creativity and imagination

One last area of 'higher mental processes', in which body and mind are obviously linked, is imagination. Using our imagination to solve problems, fantasise possible futures, rehearse presentations or see other people's point of view is clearly a valuable and highly intelligent asset. Scientists can show what is happening in the brain when we learn through imagination, and why. When we look at a situation 'through other people's eyes' we know which bits of the brain become active; and people who have damage to those bits of the brain are unable to show empathy. (Remember the Botox experiment in the previous chapter.) When we mentally rehearse a skill, brain networks specific to that skill become active and are modified as a result. You can literally improve your physical strength just by imagining yourself exercising the relevant muscles.[23] The potential of mental rehearsal in the context of practical learning has hardly begun to be tapped, yet it is clearly significant.[24]

Creativity also relies on physical hunches and promptings. Albert Einstein famously described his creative thinking in a letter to fellow mathematician Jacques Hadamard, writing that

> The words of the language, as they are written or spoken, do not seem to play *any role* in my mechanism of thought. The psychical entities which seem to serve as elements in thought are certain signs and *more or less clear images* which can be 'voluntarily' reproduced and combined ... The above mentioned elements are, in my case of visual and *some of a muscular type* ... Conventional words or other signs [presumably mathematical ones] have to be sought for laboriously only in a secondary stage, when the associative play already referred to is sufficiently established and can be reproduced at will.[25]
>
> (italics added)

People often get their best ideas when they are moving rhythmically or in very familiar ways. If you ask people 'when they get their best thoughts', they often say 'in the shower', 'walking the dog', 'doing my lengths in the pool', 'driving to work', and so on. There seems to be something about repetitive activity that puts the brain into a state conducive to creative thinking. Philosophers are known to jump up and pace about in the middle of a vigorous discussion. (My old college at Oxford even had a special woodland trail known as the Philosophers' Walk.) Creative cognition often seems to work better when it is accompanied by some kinds of physical movement. My friend John Allpress, ex-Head of Youth Player Development at the English Football Association, says that footballers are often the kinds of people who 'can only think when they are moving'. The same is often true of dancers.[26]

People actually become more creative when they are encouraged to pay attention to their bodies. Can you find the fourth word that links *eight, stick* and *skate*? These so-called 'remote associates tests' require you to access unlikely associates of each word until you find one that overlaps – and this ability is characteristic of creative solutions to more real-life problems. If you are told, 'The best way to do this task is to go with your gut feeling', you do better. Just this simple suggestion is enough to redirect your attention from more rational to more intuitive strategies. (You are also better at such creative problem-solving when you are slightly drunk, by the way, probably for the same reason.)[27] Adopting a facial expression or physical posture that is the opposite of how you are really feeling improves your creativity. The physical dissonance seems to help break mental assumptions.

*****

The embodied view opens up the radical possibility that *all* our mental functions and capabilities evolved out of this primarily somatic nature of the brain. Don Tucker makes the case strongly in his book *Mind from Body: Experience from Neural Structure*:

> The brain evolved to regulate the motivational control of actions that are carried out by the motor systems and guided by sensory evaluations of ongoing environmental events. There are no faculties of memory, conscious perception or musical appreciation that float in the mental ether, separate from the bodily functions . . . *[All] our behaviour and experience must be conceived of as elaborations of primordial systems for perceiving, evaluating and acting.* When we study the brain to look for the networks controlling cognition, we find that all of [them] are linked in one way or another to sensory systems, motor systems and/or motivational systems.[28]

What we think of as our higher kinds of intelligence – the distinctly human ones, like linguistic articulation and reasoning, mathematical and logical deduction, creativity, and social and emotional sophistication – are outgrowths of our bodily intelligence. What goes on in our bodies, and our sensitivity to those goings-on, are the roots and trunk of all the other forms of intelligence. Bodily intelligence gave birth to them, holds them firm, and continues to nurture and support them throughout life. There are no 'cognitive' regions of the brain that are not also of the body. If we took away the bits of our brain that are coordinators of bodily processes, there would be nothing left – no brain, and also no wordplay, no poetry, no algebra and no imagination.

One of the major errors of twentieth-century psychology was to suppose that there are childish ways of knowing which are outgrown, and ought to be transcended, as one grows up.

The childish ones are the bodily ones, and are to do with concrete action and experience. The grown-up ones are abstract, logical and propositional. But it is a Cartesian mistake to think that, once you have mastered logic, you don't need the body any more. Yes, new ways of thinking and knowing do emerge. But they emerge from more immediate, embodied ways, and are continuous with them. We should think of the developing mind as a tree that grows new branches, not as a spaceship whose booster rockets fall away for ever once they have done their job and are spent.

8

THE WELLING UP OF
CONSCIOUSNESS

*How thin and insecure is that little beach of white sand we
call consciousness. I've always known that in my writing it is
the dark troubled sea of which I know nothing, save its
presence, that carried me.*

Athol Fugard

It's time to tackle consciousness. Up to now I have felt the need,
every so often, to point out that much of our body-brain intelligence proceeds perfectly well without conscious awareness.
All kinds of very clever and intricate things are going on in us,
underpinning our behaviour and our experience, of which we
are not aware. Indeed, consciousness sometimes gets in the
way of our expertise, as when we become 'self-conscious' and
'awkward', losing the subtle grace of what we were doing or
saying before. A child carrying a full cup of tea is more likely to
spill it if you keep telling her to 'watch what you are doing'. Yet
awareness does seem to make a difference, often for the better.
It is certainly of the utmost importance to the Cartesian notion
of intelligence, where consciousness is seen as the theatre of the

mind, the floodlit stage on which our highest thoughts strut their stuff.[1]

So we need to ask: where does consciousness come from; when does consciousness accompany somatic activity; what is it good for; and when and why are we sometimes better off without it? Let me be clear. I am not going to attempt what philosophers like to call 'the hard problem of consciousness' – what consciousness is, in its own right, and why we have conscious experience at all. I have no idea why this 'emergent property' of the human system evolved, any more than I know why a whole lot of H_2O molecules, when they get together in a certain way, produce the emergent property called 'wet'. For the moment, I'm happy to settle for, 'Well, they just do'. The other questions are quite exciting enough for me to be going on with.

Metaphors for consciousness

As with mathematics, the place we need to start, to get an embodied handle on consciousness, is with metaphor. All attempts to discuss consciousness, even very sophisticated ones such as those of philosophers or cognitive scientists, spring from a metaphor – usually one that is very concrete. Whether they are explicit or not, these metaphors betray themselves by the language that is used to talk about consciousness. For some people, as I have just said, consciousness is a kind of brightly lit chamber or theatrical stage in the mind, and conscious experience is all that happens to be 'on-stage' at any moment. Some cognitive psychologists have associated this 'chamber of consciousness' with constructs such as Working Memory, or the Central Executive, in which case the stage becomes a 'workshop' where processes are applied to the contents.[2] This metaphor invites one to think about alternative 'dark' regions of the mental theatre where the same contents (props, actors and so on) may exist, and may carry on functioning, but out of sight of the 'viewer' – the 'I' – who is some combination of stage director and audience member.

If the actors start to behave badly, the 'I' can also act as a censor, banishing ('repressing') lewd or disgraceful performances from appearing on the stage. The Freudian image of the 'subconscious' is of dangerous material, incarcerated in a particularly dark place in the mind, which constantly has to be kept out of sight by such a censor. Sometimes this material seems to be 'alive' – to have a will of its own – constantly threatening to erupt unbidden into consciousness, grabbing the mind's steering wheel and coming out with an embarrassing 'Freudian slip'.

Alternatively, consciousness has been seen as a screen – sometimes a display panel or 'dashboard' – in the mind, on which certain information, the contents of conscious experience, can be posted. The viewer of this information might again be the first-person 'I', or, in the case of Bernard Barrs' 'global workspace theory',[3] might be other processing modules in the mind, not necessarily conscious, which can pick up the posted information and contribute their own expertise or perspective to ongoing problem-solving. The 'I' could be seen as an operational manager, watching the activity on the screen and pressing keys to determine future processing, while the chips and processors on the computer's motherboard constitute a kind of 'cognitive unconscious'. Sometimes this metaphor has been used to make a strong distinction between the objects in mental storage ('declarative memory' or 'knowledge', capable of being brought into consciousness), and the processes and programs that can be applied to that database (called 'procedural memory', usually not conscious).

Or again, conscious awareness could be seen as a roving spotlight that sequentially highlights different contents in memory. On this image, mental contents are not moved around between different locations, some of which are conscious and some not, but are illuminated *in situ*. Sometimes this illumination is seen as a form of neural activation – people often talk about different parts of the brain 'lighting up' – and the activation itself may be capable of altering the activated representation. (Animals in a forest at night behave differently when you shine your spotlight on them from the way they behave in the dark.) This metaphor, often used in the context of creativity, allows the source of illumination to be narrow and focused, or diffused, more all-encompassing, but perhaps of a lower level of intensity. A focusing mechanism allows you to 'zoom in' on a tight, bright train of thought, or you can 'space out' and look into your inner world more dimly but more synoptically.

Each of these metaphors leads us off in a different direction for thinking about consciousness. Many of them see consciousness as a special 'place' where the contents of the mind can be inspected or 'worked on'. Yet there is no evidence that there are real locations in the brain that correspond to the Central Executive or the Global Workspace, to which passive 'contents' get sent for processing, like books being recalled from the stacks of a university library and placed on a student's desk. And many of these images also rely on a mysterious agent who does the calling-up, inspecting and assembly. Who is wielding the spotlight? Who is guarding the cellar door? They cannot help us very much here, because they smuggle in the Cartesian split between Mind (the active intelligence) and Body (the passive storehouse of 'long-term memory').[4] They beg the very question we want to explore.

I am going to offer two alternative metaphors for consciousness that I will call 'unfurling' and 'welling up'. They are similar, but emphasise slightly different aspects of the underlying view I want to develop, so I shall interweave them as I go along. Let's start with *unfurling*. Imagine the growth of a plant, a fern, say, or perhaps a rose. The fern starts from an imperceptible spore in the ground, and then grows over time into a large, visible, highly differentiated mass of fronds. Let's say the spore is the original 'fertilised ovum' of an intention, an urge or a concern that is born deep down in the visceral, emotional core of the body. Over time, that motivational seed, like the fern, unfurls into a complex physiological, behavioural – and sometimes conscious – expression of that germ of an idea. Take a time-lapse film of the growing fern and speed it up, so that the whole process now takes a split second. It will be hard to spot the stages of growth. It will look as if one moment there was no fern, and then, suddenly, there is a fully formed fern-thought, or action, or emotion. In this metaphor, there is a core sense of a self-organising, dynamic process of development. There is no fully formed 'unconscious thought' that merely has to leap on to the stage, or on to which we can shine a light.

This perspective on the formation of ideas, acts and experiences was originally dubbed 'microgenesis' by Heinz Werner in 1956.[5] Werner saw the generation of a thought, for example, not as a jigsaw-like process of assembling word-meanings according to syntactic rules (the dominant cognitivist image at the time), but as a process of rapid evolution from a subcortical glimmer of meaning into an elaborated complex of sensory and motor activations across the brain as a whole, and thence back again to the muscles and the viscera of the body.[6]

Nothing in the image of unfurling yet suggests when or how, or even whether, consciousness appears. Very often the seed – the *embryonic concern*, we might call it – germinates into an

action rapidly and effortlessly without any conscious awareness. For example, we may be walking down the pavement deep
in conversation with a friend, completely oblivious to the subtle
manoeuvring of our bodies as they chart an intricate trajectory
between the oncoming pedestrians. Or we could be eating a
bowl of cornflakes while totally engrossed in a movie. If we
wanted to add consciousness to the unfurling metaphor, we
could swap the fern for the rose bush, and call the growth of the
foliage 'unconscious', and the blooming of the white flower
'conscious experience'. (This of course explains nothing by
itself, but may direct us, as metaphors often do, towards more
compelling considerations.)

To capture the experiential side of microgenesis more fully,
I will make use of the other metaphor, *welling up*; as, for
example, when we well up with emotion. Actually, welling up is
a deeply embodied metaphor; many of our experiences of the
body have this developmental quality. A sneeze and an orgasm
are capable of being noticed as they gather muscular strength
and patterning. But so do a bubble of laughter (that we may
struggle to contain) and the feeling of being moved to tears.
Some of our more psychologically tinged experiences also
display at least aspects of their unfurling in consciousness. They
start, if we are being attentive, vague, hazy and slight, and grow
in shape, definition and urgency.

Most of us are familiar with this experience of 'welling up'
– occasionally. But my more radical suggestion is that these
special moments are actually prototypical of our conscious
experience as a whole. All our thoughts and sensations well up
from visceral and unconscious origins in the same way, though
we may not notice the unfurling. Benny Shanon sums up this
integrated, embodied view of the mind, using a different set of
metaphorical terms, like this:

Fully-fledged cognitive patterns are generated out of a substrate which is qualitatively different... [These substrates] are not cognitive or mental in the sense that manifest patterns are. The process of this generation is one of differentiation and fixation, whereby patterns with well-defined structural properties are created out of a substrate lacking such properties... I call the process of this generation *crystallization*, and I maintain that this is one of the fundamental operational features of the human cognitive system... *[Conscious] structures are not the basis for mental activity but rather the products of such activity.*[7]

That last sentence captures more formally the essence of what I am trying to say. Shanon's 'substrate' of conscious thought is the bodily/emotional core of meanings, values and intentions that I have been talking about. These somatic stirrings are as different from conscious thoughts as the fibrous roots of a fern are from the exuberant intricacy of its fronds.

The welling up of gesture and thought

Everything is the way it is because it got that way.

D'Arcy Thompson

Let me bring some evidence to bear on this metaphorical proposition. If the 'unfurling fern' view is right, then we should see the original seed differentiating into different 'branches' as the unfurling takes place. For example, something we might eventually say, and all the non-verbal gestures that accompany it, may well stem from a common embryonic concern, and carry different aspects of the original meaning or intention. Our 'utterance' is the whole fern, not just one particular frond. Over many years of painstaking research, psycholinguist David McNeill at the University of Chicago has demonstrated that this is exactly what happens when we are trying to communicate.

His research focuses on the relationship between what we say – for example, when we are describing a cartoon we have just watched to a third party – and the hand gestures that spontaneously accompany the speech. Through detailed analysis of videotapes, McNeill and his collaborators have discovered that speech and gesture do indeed emerge from the same root, and carry complementary aspects of the meaning we want to convey. For example, describing a scene, one observer said, 'Sylvester was in the Bird Watchers' Society building, and Tweetie was in the Broken Arms Hotel . . .' As she referred to Sylvester she gestured to her right-hand side, and as she referred to Tweetie she gestured to her left, indicating that the two locations were on opposite sides of the street. A few seconds later, she reported that 'He then ran across the street', and gestured to her right as she did so – indicating that it was Sylvester who crossed the street, not Tweetie. Speech and gesture were woven together seamlessly – and unconsciously – in order to resolve the potential ambiguity of the pronoun 'he'.

We might imagine – if we could slow the 'movie' of our own experience down sufficiently – that we might catch the germ of the desire to communicate something as it begins to stir deep in the body-brain. It might involve a need for approval or a desire to impress – to want to be a 'good subject', in McNeill's experiment – or a wish to convey involvement and amusement in the cartoon, or a dozen other intentions. When I say to my wife, 'I think the front lawn needs cutting', I can trace the source of that casual comment back to a small archaic tremor of potential concern about being disapproved of by meticulous neighbours. When she says to me, 'Shall we go for a walk if it's nice tomorrow?' I can, I fancy, hear a faint echo of anxiety about my health and the sedentary nature of my work.

The broad structure of these fronds of communication is determined by the genetic programmes that shape our bodies. The actual expression of these genetic guidelines, though, is powerfully modulated by the accidents of our experience. So the fine details of how an embryonic intention unfurls are heavily experience-dependent. Synaptic connections are changed and chemical responses throughout the body altered by learning, so the pathways along which meanings unfold are individual and variable. At the risk of creating metaphorical overload: a gathering tide of activation flooding through the body is shaped by the channels and contours that thousands of previous tides have sculpted and left behind.

As we saw in earlier chapters, some of these ripples will cause visceral and autonomic ramifications: blood pressure and heart rate might go up, respiration volume might become shallower, background processes of digestion might be inhibited, hormones such as adrenaline or cortisol might be released into the bloodstream. All of these changes will be signalled to the brain and will alter levels of arousal, as well as the focus of attention (what we are on the lookout for) and the memory contents that are recruited into the unfolding meaning. The brain alerts muscle groups for the kinds of action that might be required, and facial expression and patterns of body tension are altered accordingly. Some of these muscular changes might involve coordinated patterning of throat, mouth and lips, at the same time as the lungs are expelling air in synchronised bursts – resulting in the utterance of a sentence or a sigh. And some more muscular activity might result in simultaneous movements in shoulders and arms that produce gestures which disambiguate or augment the meaning of the utterance.

We might call these the four major fronds of the developing 'fern' of experience. One branch elaborates internal, 'interoceptive' body states: visceral, hormonal, immunological and neural. A second alters the direction and acuity of incoming sensation, via modulation of the 'exteroceptive' perceptual systems. A third branch begins to ready muscle groups for direct action. And the fourth branch may head in the direction of linguistic and other kinds of symbolic output such as gestures. And, as the fronds of the fern remain connected at the heart of the plant, so all of these remain functionally looped together in the body-brain.

For example, suppose you are one of the participants in McNeill's study. The seed of your desire to retell the story of the cartoon fires up memories and their emotional concomitants. These then cause your voice quality to intensify, a chuckle to form in your chest and the corners of your mouth to arch upwards. At the same time, the internal stirrings begin to activate appropriate patterns of words. The right muscles are recruited to shape your voice box and expel air to make the right sounds. And you establish eye contact with the listener in order to judge whether you are being successful in conveying the meaning and the feeling that you intend. Your arms and hands begin to form gestures that carry other aspects of the felt meaning you want to convey. A whole-body state fans out from the originating impulse, unfolding into an intricate, evanescent pattern of neural, muscular and hormonal events.

It has been shown that children's arm movements can actually convey a more creative part of their overall understanding than their tongues or pens (or iPad screens). As they talk, children naturally gesticulate, and their gestures carry part of the meaning of what they are trying to say. One classic test of cognitive development is children's understanding of what is called 'conservation of volume'. If water in a tall thin beaker is poured carefully into a short fat one, the appearance of the water is different but its volume stays the same. Younger children are unable to escape the appearance and say that the 'amount' of water has changed.

If children are made to sit on their hands while they are talking about this phenomenon, they appear to be at this lower level of understanding. The unfurling of their understanding is hampered when they are deprived of gesture. Careful analysis has shown that children who, judging by their verbal explanations, are not quite at the higher level nevertheless show through their gestures that they are in fact aware of the conservation of volume. (They might cup their hands and move them from right to left, as if physically transferring the water, without changing the size of the 'cup', for example.) Susan Goldin-Meadow, the author of these studies, suggests that gesture is able to express more creative insight than the more conservative verbal 'frond' is willing to venture. What we say is judged more harshly than the way we move our hands, so gesture takes advantage of the slacker criteria to try out newer, more tentative ways of looking at the world.[8]

This close association between language and gesture also receives support from studies of brain organisation. As we saw earlier, Broca's area – a brain region that has a lot to do with language production – lies next to the part of the brain that controls hand movements, suggesting that spoken language itself piggybacked on an earlier system of communication through gestures. According to one suggestion, our early ancestors first developed a kind of manual sign language, which gradually became augmented with characteristic vocalisations. As the vocal tract evolved to become more sophisticated, and the muscular control of tongue, mouth cavity, throat and breathing became more refined, so speech began to run ahead of gesture, and eventually became the dominant partner.

Espaliered experience

As the germ of an idea evolves into a linguistic utterance (or a written or 'signed' sentence), all kinds of syntactic and semantic considerations come into play. Through spreading activation in the loops of the body-mind, the unfurling idea starts to recruit candidate words and syntactic frames to carry the intended meaning. But it could be that no readily available words or frames are capable of accurately conveying the underlying intention, so, if the utterance is to proceed, some of the nuances and subtleties of the meaning may be lost in transcription, and what eventually comes out is only an approximation – perhaps a crude approximation – to what was intended. The horticultural term *espalier* refers to the process of training a plant – often a fruit tree – into a particular shape as it grows, through selectively pruning offshoots and tying other shoots to a frame that directs their growth. Language acts as a kind of espalier for the meanings that are germinating within the body-mind.

One kind of 'training' that happens to the embryonic intention is so common that it is rarely noticed. It involves the linguistic necessity, present in many languages, to insert references to a kind of 'self'.[9] Instead of just experiencing a thought, perhaps 'Is the cat in the bedroom?', we say 'I wonder if the cat is in the bedroom'. Added to the thought is a somewhat gratuitous wonderer – just as when we say 'It is raining', there is no 'it' that is doing something called raining. 'I', like 'it', is a convention. (Or: 'I' is like a contour line on a map. It is a symbol that has no referent in the real world. It is not necessary to worry about tripping over the contour lines as you climb the hill.)[10]

Over time, as a child learns to use the dummy word 'I' correctly and fluently – not just 'I fell' or 'I ate', but the more puzzling 'I saw', 'I tried', 'I decided' and so on – it comes to seem that 'I' does indeed betoken some kind of ever-present ghostly observer, instigator, manager, editor or narrator lurking behind appearances. We get used to adding this ghostly espalier to each meaning as it unfurls. Eventually it appears self-evident to us that there is a (real, albeit spectral) observer who is capable of watching thoughts and experiences as they appear in and disappear from consciousness, and a real inner agent who does all the intelligent thinking and deciding. The Chief Executive of the mind, so central to the Cartesian view, turns out to be a linguistic convention rather than a potent force. *The actual business of thinking is embedded in the process of unfurling; it is a function of the whole dynamic body-brain system, not of any mysterious Fat Controller of the mind.*

But 'I' is not only redundant (as an internal agent); it also has the ability to make trouble. In daily life, the assumption that 'I' refers to a real inner entity can make self-critical judgements feel both more serious and more 'sticky'. For example, in a linguistic construction, such as 'I tried but failed', there is apparently some *thing* (or some *one*) to which the judgement can adhere, and which is therefore obliged to feel culpable for the 'failure'. Unnecessary depression and anxiety may ensue. Various kinds of therapeutic procedure, such as Mindfulness-Based Stress Reduction, encourage people to observe the arising of thoughts such as 'I'm a failure' simply as a welling-up rather than as a valid judgement.[11]

Habits of attention

The process of welling up can take quite different time courses. Sometimes it takes only a fraction of a second, in which case it is very difficult to catch the unfurling as it happens. And sometimes the development of a germ of an idea into a communicable train of thought takes seconds, or minutes, or even longer. We may have to 'um' and 'er' while we wait for the *mot juste* to come to mind. Sometimes the unfurling is blocked, as when our brain refuses to come up with the name of a dear friend when we are having a 'senior moment'. In many stories of creative insight, the solution to a problem hangs elusively just out of our grasp for months or years, until exactly the right set of triggers comes together and the answer is finally 'propelled into consciousness'.[12]

However, overlaid on these different time-scales there may be habits of attention that make us more or less sensitive to the unfolding dynamics within. We may develop a generalised habit of not paying attention to the early stages of the unfurling so that, whatever its intrinsic time-course, we do not become consciously aware of what is welling up until late in its development. We come habitually to notice aspects of experience that are already well formed and elaborated, but do not notice their hazier precursors. Thus, instead of noticing the gradual clarification and differentiation of a thought or a feeling, we experience our own experience in terms of a step-wise distinction between things that are not conscious and those that are. They appear to 'pop into' our minds – or even, in a magnificent sleight of hand, to spring, fully formed, out of the mouth of the 'inner I'. Thus – referring back to the earlier discussion about different metaphors for consciousness – what looks like a structural separation between conscious and unconscious can actually be a reflection of an acquired cognitive habit. We see a sharp distinction, but only because we are inattentive to the gradient that leads from unconscious to conscious.

We might also develop a habit of speeding up the unfurling process itself by drawing on the body-mind's power to anticipate its own states. We know that our embodied system is capable of registering regularities in its own processing, and is therefore able to predict (with greater or lesser accuracy and/or confidence) what mental states are about to happen, on the basis of what mental states are already happening.[13] This ability to anticipate states of our own mind makes it possible for us to take short cuts in unfurling, and leap to conclusions about what probably or usually follows the current state. To use our earlier prosaic example, a glimpse of what looks like Timmy's tail disappearing round the corner of the house leads us to expect, were we to run and peek round the corner, that we will see the whole cat; but if we are lazy, or it really doesn't matter that much, we can just assume it is (the whole) Timmy without bothering to check.

When we have to respond fast, taking such short cuts can be advantageous: it can even save our lives. But if leaping to conclusions becomes habitual, we are likely to miss detail and novelty. We construct our own world based not on the unprecedented particularities of the moment, but on what is normal and expected. Thus we can come to see in terms of faint, familiar stereotypes and generalisations rather than the vivid, complex individuality of what is actually present. (We experience a shadowy stand-in for Timmy. Had we checked, we might have seen that it was in fact an entirely different cat for which, we have just read in the local paper, a distraught owner has offered a substantial reward.) In effect we are trading vitality and inquisitiveness for normality and predictability.

Going too far in this labour-saving, top-down direction doesn't just make life duller; it obviously incurs risks and costs. We might try to make the world conform to our expectations, and thus persist in applying methods of thinking and acting that worked once but are not, in a new situation, appropriate or effective. (Applicants for jobs at Google are often asked if they have a track record of success in their field. Those who boldly say yes are unlikely to be hired, because experience has taught Google that such self-confidence often leads people to try to replicate those successes by forcing new predicaments to fit old patterns; whereas Google is interested in people who can think from scratch and 'flounder intelligently' in the face of quite new challenges.)[14]

Language can certainly exacerbate this problem. There are many studies showing how a verbal label often leads to a kind of 'functional fixedness' in which alternative ways of looking at or categorising an object are rendered invisible by the label. In one classic study, researchers showed people ambiguous pictures with one of two suggestive verbal labels, e.g. 'dumb-bells' vs. 'spectacles'. When participants were asked to draw the shapes from memory, it was found that the labels markedly skewed what they thought they had seen (as in Figure 10).[15] This is one way in which creativity is reduced by leaping to conclusions. Creativity also suffers in other ways. We can become deaf to our own inklings and hunches, which recent research has shown are vital aspects of creativity. Being able to access and tolerate what some researchers refer to as 'low ego-control' or 'low arousal' mental states – those that are uncertain, provisional, ambiguous or vague – is demonstrably conducive to creative insight. These hazier experiences are at an earlier stage in their unfurling. They may be carrying the undifferentiated seeds of a new thought. Rushing the unfurling, forcing the hazy intuition into more familiar tracks, may well sacrifice the latent insight.[16]

The process of unfurling can be slowed down as well as speeded up. As we saw earlier, the process of checking candidate actions or utterances for accuracy and completeness may be subject to strategic control by those inhibitory frontal lobes. We can allow the 'stream of consciousness' to flow, or we can monitor and edit more carefully. When speaking a foreign

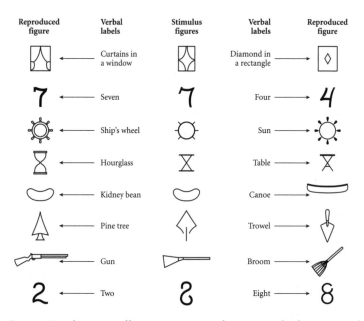

Reproduced figure	Verbal labels	Stimulus figures	Verbal labels	Reproduced figure
	Curtains in a window		Diamond in a rectangle	
	Seven		Four	
	Ship's wheel		Sun	
	Hourglass		Table	
	Kidney bean		Canoe	
	Pine tree		Trowel	
	Gun		Broom	
	Two		Eight	

Fig. 10 How language affects memory. Based on Carmichael, Hogan and Walter's 1937 experiment.

language, for example, some people may dive in and 'give it their best shot' (especially in convivial company or after a drink or two), while on other occasions (or if they are temperamentally more cautious) they may self-monitor to the point of becoming tongue-tied.[17] When a possible thing to say is being held back for checking, the motor programmes for producing the speech can be run 'off-line'. That is, the incipient muscle movements for saying it can be run in a muted way that produces internal 'thought versions' which can be checked and edited. The eventual utterance will likely be more correct, but there is also a strong possibility that the conversation will have moved on, and the well-honed comment just won't be relevant (or funny) any more.

The emergence of consciousness

As I have said, none of this unfurling is originally or necessarily conscious. Often the major frond involves only unreflective action. But, in the course of their welling up, different kinds of somatic events and patterns *may* be produced that do cause a conscious experience to occur. Such experiences come in a variety of forms. One is the perceptual world of sights, sounds and so on, which we interpret as a 360-degree wraparound backdrop to all our actions. Then there are more or less clear bodily sensations, feelings, emotions and moods. There are expectations, premonitions and intentions: feelings of readiness or anticipation; the feeling that we know something before we can recall it. There is a whole range of inklings, hunches, promptings and other kinds of intuition. There are verbal or symbolic thoughts, as well as internal sensory and muscular images. And there are 'memories' (images that come tagged as recollections of past events) and 'fantasies' (images that come tagged as possible or probable future events).

Consciousness varies not just in contents but in its quality. If what is welling up starts to turn into a *conscious* welling up, it can do so not only in varying forms but with varying degrees of clarity and intensity. Sometimes there might only be glimmerings of awareness: the haziest and faintest of apprehensions. Sometimes we are not even sure if we are feeling anything. Sometimes we know that we are feeling something, but can't yet pin it down. Sometimes these inner works-in-progress unfurl further, so that what was hazy begins to take a clearer, more differentiated form. Yet sometimes insights, emotions or thoughts burst into consciousness with the utmost force, precision and/or certainty. And as I discussed earlier, these stages of unfurling can be further muddied by our own habits of attention. Some people can tune into the faintest signals from their bodies; others don't seem able to feel it even when they are, to anyone else, visibly fearful or angry.

*****

All of these varieties of conscious experience constitute what we might call complementary 'ways of knowing'. Each of them reflects something of what is going on in the interior. Each of them is useful, and each of them has flaws. However, on the Cartesian model, only disciplined trains of thought count as proper 'thinking', and only the conclusions of such trains, their logically arrived at termini, as 'knowledge'. But we know all too well how stupid logic can be. After Queen Elizabeth II had wondered aloud how economists could have failed to foresee the credit crunch, an eminent group of them sent her an explanation. They wrote: 'In summary, Your Majesty, the failure to foresee the timing, extent and severity of the crisis and to head it off, while it had many causes, was principally a failure of the collective imagination of many bright people, both in this country and internationally, to understand the risks to the system as a whole.'[18] Imagination failed because the possibility of the crash was not part of the economic logic they were using. Logic can handle only a small number of well-defined factors at a time, so to think logically, we have to shrink the real, messy, confusing world to accommodate that constraint – and in doing so, often lose much of its vital richness and significance. It's Garbage In, Garbage Out, however logical you are.

If logic has weaknesses as well as strengths, the other ways of knowing have a value that counterbalances their own short-comings. Physical promptings and intuitions, for example, are not always reliable, but there are times when they can inform our decision for the better. 'By the pricking of my thumbs, something wicked this way comes,' intoned the witch in *Macbeth*, and we now know that our knowledge banks do sometimes speak to us in somatic ways. International financier George Soros takes account of his lower back pain when making a financial decision. Remember it is the physical arousal of the skin that is the first signal of learning in Damasio's Iowa gambling task. Intuition and fantasy are of proven value and importance in all kinds of intelligent endeavours, from Nobel Prize-winning science to poetry-writing to preparing for elite sports. (Jack Nicklaus famously said: 'When I hit a golf ball, [first] I go to the movies in my head'.)[19]

When is consciousness?

The different ways in which the body's neurochemical activity may eventuate in conscious awareness are under intensive investigation, but the question I want to focus on now is a more general one: *when* does any of this dark and silent internal activity, whatever its eventual form, become associated with consciousness? It is widely agreed in neuroscience that conscious awareness – whatever it is and whatever it is for – emerges alongside complex neurochemical states of biological organisms like us.[20] No one yet knows exactly what character-ises those particular states, nor exactly which bits of the body-brain are crucial for consciousness. We can be sure, however, that they are complicated, integrated and distributed. There are strong suspicions that they involve regions of the brain like the anterior insula, the anterior cingulate, and areas of the prefrontal cortex, where all the various neurochemical loops come together and talk to each other; but other, more widely distrib-uted, circuitry is almost certainly involved as well.

However, fools rush in, so I shall take a chance and say that, to give a rough approximation, consciousness seems to emerge when some combination of five conditions are present. The first is *intensity*. Consciousness accompanies external events that are sufficiently abrupt or intense: for example, the sudden ringing of an alarm bell, the flashing of a bright light or a piercing pain. The second is *persistence*. Intensity seems to interact with the persistence of a stimulus: less intense stimuli often become conscious if they persist for more than around half a second. Softer events can build up to the requisite level of activity for consciousness if they are maintained.[21] (At longer levels of persistence, consciousness may fade again through habituation, of course.)

The third condition is *reverberation*. Persistence may occur not because of the physical continuation of an external event but because of conditions within the brain that allow activation to reverberate, for example round a well-worn neural circuit where resistance is low and activation can, so to speak, keep rekindling itself. It is argued, for example, that we can easily retain a sensible sentence in mind because its elements 'fit together' and create such a reverberating circuit, while a list of random numbers needs to be continually rehearsed or refreshed if its elements are not to 'fade away' and become inaccessible.[22]

The fourth condition for consciousness is *significance*. Consciousness seems to be attracted by experiences that are of personal significance. These may be threats to physical well-being or survival, or to possessions or personal attributes with which one identifies. Even events weak in physical intensity gain access to consciousness under these conditions: a creaking floorboard in a sleeping household; a disapproving expression on one face in an otherwise positive audience. Researchers such as Antonio Damasio, and Gerald Edelman and Giulio Tononi, have suggested that self-related events attract consciousness because they connect with a constantly active representation of the 'core self' in the brain, and thus become part of a massively reverberatory circuit. This isn't a fixed, structural circuit; it functionally connects whatever aspects of internal neurochemical activity happen to be locked into the 'core self' at that moment. Thus the ingredients of this core self-circuitry are constantly changing as, for example, some concerns are dealt with and drop out and new ones arise.[23]

And the fifth and final condition that is likely to bring about consciousness is *checking*. Activity often generates conscious experience when it is being internally checked or inhibited. As we have seen, the prefrontal cortex exerts all kinds of inhibitory controls over what is going on elsewhere in the body-brain. This is, in terms of the internal neurochemical economy, intensely energy-consuming – and it also creates dammed 'pools' of blocked activation that can build up to higher levels of intensity. So there is a strong (but not invariable) association between these inhibitory stockades and corrals, and the emergence of linked conscious experience.[24] This occurs, for example, when we are holding back candidate courses of action, such as a contribution to a conversation so that the cost of a (social) error can be assessed; or when 'struggling with temptation'.

Acts of disciplined thinking tend to become conscious because they require a narrow defile of activation to be maintained against competing possibilities and distractions. (That's why 'working memory' is strongly associated with consciousness: not because 'that is a place where the light of consciousness shines brightly', but because the inhibitory effort required to keep tight control of internal events creates the kind of intense, corralled activation that is associated with the appearance of consciousness.) Obviously, whatever is currently connected to the 'core self' is likely to be subjected to more extensive 'security checks'. (Dispositionally anxious people may be constantly on the lookout for remote threat possibilities, and thus become highly self-conscious and 'inhibited'.)[25]

This last point is worth stressing. Conscious rational thought is not a different kind of event from feeling or seeing. It too has its roots in the embryonic concerns of the body-brain; it doesn't come from different stock. Where it differs is in the intensity of activation it attracts, as it unfurls and emerges, by virtue of the degree of neurochemical constraining that it requires. Reasons and arguments well up from the dark depths of the body-brain just as emotions, intuitions and images do. And we often talk to ourselves when there is an underlying tangle of intentions, affordances and opportunities – when the way forward is unclear or conflicted. Such predicaments tax the body-brain's inherent problem-solving and conflict-resolving procedures, and a great deal of neurochemical excitation and inhibition gathers around the conflict. Out of this kind of biological situation, where action is slowed down or blocked completely, emerges a swathe of consciousness-attracting activity, some of which takes the form of inner speech – i.e. thoughts.

Overall – if you can stand another metaphor – the body-brain functions rather like a pinball machine. Sometimes an event occurs and, especially if it falls within an area of expertise, the system responds quickly and silently. Other times, when the resolution of Needs, Deeds and Seeds is less well honed, or where the costs of error are estimated to be great, there is much more protracted rattling around between different possibilities, and the brain's electronic bumpers keep the ball ricocheting around, accompanied by the flashing lights and sounds of consciousness. At such times the *son et lumière* reflects the complicated activity but it does not *control* it.

189

I wells up too

Where does this leave introspection – the idea that the mind is a clock in a glass case: a mechanism that the owner can inspect (should she have a mind to)? Where the earlier metaphors for consciousness invited, indeed required, the idea of this ethereal agent (the CEO at her well-lit desk; the wielder of the spotlight), the images of unfurling and welling up do not. The sense of the inner author, instigator or observer is just another frond of the unfurling fern. Both 'I' and 'thought' emerge, in the moment, as aspects of the same upwelling experience.[26] Whatever ideas or experiences come into consciousness, they are, as Shanon says, *products* of this intricate unfurling, not a direct inspection of it. We never see our own process. Never. The workings of the body-brain are entirely dark and eternally silent. But sometimes out of the gates of the forbidden factory rumble trucks carrying audible and visible goods. The *sense* of being able to inspect ourselves is 'real': as real as anything else that wells up and takes transiently conscious form. But the idea that this sense refers to a real ability to lift up the hood and look inside ourselves – that's just another idea, and not an accurate one. It's a unicorn.[27]

The idea that consciousness is 'for' some separate inner 'I' to look at runs deep in our culture – and certainly in psychology. To quote just one example, from a recent paper: 'Even though intuitive processes themselves remain unconscious, they produce outcomes such as intuitions or gut feelings that *we can consciously attend to . . . and use in our judgements*.'[28] Familiar enough, and seemingly innocuous – yet who or what is this 'we' that is attending, being informed and making judgements? Both everyday and academic talk constantly smuggle in the idea of the ghostly agent – the mind – that is clearly other than the workings of the body-brain itself. The idea that thinking, and the feeling that there is an inner someone who is 'doing' the thinking, co-arise when the body-brain finds itself in certain kinds of state is hard to keep hold of, especially once we start talking!

*****

This view of the self also fits with a wealth of research that shows, as the University of Virginia's Timothy Wilson puts it, that we are 'strangers to ourselves'.[29] We know ourselves, it turns out, only through two routes. The first is by observing our own behaviour, and making inferences about what the internal causes or reasons must, or might, have been. Occasionally we admit to this inferential process – 'I don't know what came over me. I must be more stressed than I'd realised' – but mostly we do not. The research shows that we confabulate reasons for our behaviour all the time, and then treat them as if they were direct readouts from the interior – but they are often wrong. Show people an array of identical women's tights and ask them which they prefer. (Control for other variables like the lighting and where exactly people are standing.) You'll find that the majority of them will pick out the pair on the right. Ask them why they chose that pair and they will give you all kinds of reasons. What nobody ever says is 'I chose the pair on the right because that is what I tend to do.'

It turns out that this inferred self-knowledge isn't very good. Other people who know you reasonably well will predict your behaviour better than you will. Your own estimate of how generously you will give to charity, for example, is, well, generous. If you want to know yourself better, ask a friend. And that's the second source of self-knowledge: believing what other people say we are like. In adulthood, if we have reasonably good and intelligent friends, what they say about us is likely to be pretty accurate. But it was not always thus. In childhood, other people can be very keen to tell us who we are, and how we ought to be. They may lock us into a cage of attributions based on our gender, our skin colour, or even just on their need for us to be predictable and untroublesome. But those attributions espalier our inner workings, potentially misdirecting our actions and distorting our consciousness.

9

THE AUGMENTED BODY

Your self does not end where your flesh ends, but suffuses and blends with the world, including other beings.

Sandra and Matthew Blakeslee[1]

We have been gradually expanding our understanding of the body: what it is, and how intelligent it is. And we have been working our way up from the 'sub-personal' level of molecules and tiny electrical sparks, through the interwoven systems of the body, to the 'personal' level of conscious thoughts and the more prototypical forms of human intelligence – talking, decision-making, problem-solving and creativity. But you'll remember that, when we were talking about the nature of human beings as 'complex adaptive dynamic systems', way back in Chapter 3, we saw that, not only are we composed of Sub-Systems, but we are ourselves Sub-Systems within a whole array of nested Super-Systems. To comprehend ourselves correctly, we have to include the 'supra-personal' level as well.

The envelope of our skin (and all the specialised kinds of touch-receptors we have evolved, like eyes and ears) is not a boundary or a barrier but the place where we are joined to the world. To understand an isolated body on its own makes no more sense than trying to understand the heart without looking at its place within the life of the body. A heart in a steel dish is profoundly different from the same heart *in situ*. We saw that the healthy heart is in constant resonance with all its fellow organs and substances – and we human beings, through our bodies, are in continual resonance with the flux of energy, activity and information that surrounds and batters us.

The fact that much of this interaction occurs unconsciously makes it no less real. But we have acquired the cultural habit of overvaluing those aspects of ourselves that are conscious and explicit, and this makes us neglect this constant reverberation with our surroundings. Some of these interactions are physical, and often pretty obvious. We die if we stop exchanging gases with the world. If we lose the sense of touch, whether the comfort of a caress or the pain of an injury, we are badly disabled. But many of these interactions are social, and often more subtle. We react to the expression in someone else's eyes without knowing it. We are sensitive to people's smells of which we are not aware. In this chapter we will explore some of this tissue of connections with the environment.

My-space

Being tickled makes most of us squirm – but so can a pretend tickle that makes no contact with the skin. Waggling your fingers in a threatening manner close to a child's body can elicit the same kind of squirming and giggling as a real tickle does. That is because there is, around your solid body, an invisible bubble which scientists call the 'peripersonal space'. It is the area in which, without shifting your overall position in space, you could grab, stroke, kick or butt something if it came your way.[2] It is the zone of direct interaction with the physical world: the principal arena in which we use our limbs to latch on to things we desire or that interest us, or to fight off or bat away things that are noxious or threatening. As we saw in Chapter 5, objects that enter the peripersonal zone (such as a rattle offered to a baby) are processed by the brain in a different way from those that are outside it. Their graspability, for example, becomes a much more salient trait, and the motor areas in the brain that underwrite grasping are automatically primed for action. The same object, seen outside the periper-sonal zone, is processed in terms more of its own characteristic sensory features than of the physical actions needed to grab hold of it.

In 1994, Michael Graziano and Charlie Gross at Princeton were exploring the behaviour of cells in the monkey brain that seemed to respond to both touch and vision. The cells would fire off when, for example, the back of the monkey's hand was stroked; but, Graziano and Gross discovered, the same cells would fire equally strongly if the monkey saw the experiment-er's hand get close to the back of its hand without actually touching it. It looked as if the visual clue was enough to prime the brain to expect a certain kind of touch sensation. In effect, the monkey's sense of touch extended out from its body like an 'aura'. Of course – unlike some parapsychological claims for the reality of auras – the effect only occurs if the monkey can see the approaching stimulus: that is, if it 'knows' that something has entered its peripersonal space. It has to be close – in the monkey's case, around eight inches away – before the effect begins, and the cells' responses get stronger as the stimulus approaches closer and closer. The effective body extends out beyond the solid pillar of meat that we usually call 'the body'.[3]

Graziano and Gross also found that many of these cells are not where the traditional view would have them be: tucked away towards the back of the brain in the parietal cortex, where information from the senses is known to come together and combine into multisensory representations. No, some of these cells were up front, in the premotor cortex. Here we have another clear example of the interwoven, sensorimotor view of the brain we looked at in Chapter 4. The whole point of being able to predict that something is about to brush your hand is to be able to adjust your hand to grasp it, avoid it or just let it arrive (depending on what you are up to, and what you think it is). The brain is designed to let perceptual predictions and action readiness resonate together virtually instantaneously, and it is smart of it to do so.

However, this inbuilt tendency to lock our attention on to objects that are near our hands – and to process them more quickly and more fully – does have a downside. It is harder for us to shift our attention away from such objects to something else. Give people a task where they have to look at a computer screen and switch focus quickly from one blip on the screen to a second one that happens very soon after. If you tell them to put their hands either side of the screen, rather than in their laps, their ability to switch is slower. The researchers suggested that 'objects that are near the hands are likely candidates for physical manipulation [such as tools or food] ... In those circumstances, extended analyses of objects near the hand may facilitate the production of accurate movements'. But it hinders your ability to deal with interruptions.[4]

A striking demonstration of the different ways the brain treats the peripersonal zone and the space beyond has been reported by Italian neurologist Anna Berti. One of her patients had a brain disorder that made her seem oblivious to the left side of her vision – but only for things that were within reaching distance. When she was asked to point to the middle of a horizontal stick close to her, she pointed way off to the right-hand end. But when the stick was moved further away, and she was asked to indicate the middle using a laser pointer, she got it more or less right.[5] In everyday language we mark this shift by talking about 'this book' (within grasping distance) as opposed to 'that book' (further away). It would sound odd in English to say, 'I'm sorry to trouble you, but could you possibly pass me this book.'

The functional zone around the body isn't fixed, though: far from it. When it needs to, your brain is able to adjust the size and shape of this My-space depending on what you happen to be wearing, driving or using at the time. When we pay someone an old-fashioned compliment and say 'That hat becomes you', we are speaking the literal truth. In the days when it was common for women to wear hats with wide brims and large feathers, their body schemas would quickly adjust so that they could navigate narrow doorways and low beams without mishap, and without thinking. A renowned neurologist of the day, Sir Henry Head, wrote, 'A woman's power of localisation may extend to the feather in her hat,' and more generally, 'Anything which participates in the conscious movement of our bodies is added to the model of ourselves and becomes part of the [body schema].' Clowns on stilts do the same thing. Experienced drivers of huge articulated trucks know without thinking where the back wheels are to the nearest centimetre. In fact they will usually talk about 'my back wheels', not 'the back wheels of the lorry I happen to be driving', as if the truck had indeed become incorporated into their sense of their own body. My-space is also elastic in terms of how much the things in it, or outside it, matter to you. Desirable objects towards the outer limit of the reachable space are seen as nearer than undesirable or unpleasant ones. That is to say, the peripersonal zone is bigger for things you are keen on.[6]

Tools, as far as the brain is concerned, quickly become incorporated into the body. When I am using a pencil or a tennis racquet, my brain automatically adjusts the peripersonal space to include the tool. A blind person's stick literally becomes part of their body map. When people have been using a 'grabber' to pick up litter, they judge the distance between elbow and fingertips on the hand which had been using the tool as longer than on the other arm. We literally feel as if our own arm has been elongated. Intriguingly, when Anna Berti's patient was asked to point to the middle of the far-away rod using a long stick rather than the laser pointer, her distorted view of space re-emerged. With the physical pointer, her brain now 'saw' the rod as within the peripersonal space – and so the deficit associated with that near space came back.

In a detailed study of the malleability of these body maps, Professor Atsushi Iriki at the RIKEN Brain Institute in Tokyo has trained macaque monkeys to learn how to use a small rake to drag in raisins – a favourite treat – that are beyond their unaided reach. To see how the brain responded to this new skill, Iriki implanted tiny electrodes in the posterior parietal cortex of the rake-wielding monkeys and measured what are called the *receptive fields* of individual cells while they were using the rake. The receptive field is the range of physical locations that will elicit a response from a particular cell. Without the rake, the cell's receptive fields covered just one hand, for example, and the small zone of peripersonal space around it. But when the monkey was using the rake, the same cell now included the rake itself, and the whole wider area that could be reached with the rake. Interestingly, when the monkeys were just passively holding the rake, but not actively using it, the receptive fields quite quickly shrank back to their smaller size. For us virtuoso tool-using human beings, the extended body map lasts longer after we have put down the tool: about 15 minutes.[7]

Avatars provide an interesting tool for exploring just how malleable these body maps are. Back in the 1980s, virtual reality (VR) pioneer Jaron Lanier and his colleagues were playing with the kinds of VR avatars that the brain would accept. An avatar is the way your own body looks to you in the VR environment. When you look down, or look in a virtual mirror, that is what you see. You can go as yourself or in fancy dress. Sensors attached to your limbs relay your muscle activity to the computer, and those signals are used to drive the movements of the avatar image.

After a while, Lanier's colleague Ann Lasko got bored with programming avatars with two legs and two arms sticking out of a human-like torso, so she added six large lobster legs to the chest of Lanier's avatar to see how he would cope. Would his brain be able to learn how to control the legs? In fact, she had programmed in some very tricky rules that related what Lanier could do with his real body and how the virtual lobster legs would behave. A subtle combination of the angles of his left wrist, right knee and right shoulder, for example, would make the bottom left lobster leg flex in a particular way. As Sandra and Matthew Blakeslee say, in describing this experiment, these patterns were 'much too complex and subtle for his rational mind to grasp.'[8] Yet, after some time in the VR environment (as Lobster Man), and without understanding what he was doing, Lanier became able to control the imaginary legs to a remarkable degree. 'After a bit of practice I was able to move around and make the extra "arms" wag individually and make patterns of motion,' said Lanier. 'I was actually controlling them. It was a really interesting feeling.' I bet it was!

Lanier and his team went on to design an avatar that had a tentacle sticking out of its belly button, which people could learn to wiggle around in the same way as the lobster arms and legs. But this time they combined that trick with another one. If you get people to wear special gloves that vibrate in a certain way, you can create the feeling of a physical sensation out in mid-air, in between their two hands. This in itself, Lanier says, is a truly strange experience. But now, if you rig things right, you can get that phantom feeling to coincide, in space, with the tip of the imaginary tentacle – and it feels just like a throbbing in a real part of your own body!

You can even co-opt parts of other people's bodies into your own. Imagine you are sitting at a table with your left arm stretched out in front of you, but hidden from your view by a cover. Nearby, and visible to you, is someone else's arm stretched out. A third person uses one of his hands to stroke and tap your arm – the one you can't see – in a distinctive pattern and rhythm, and at the same time uses his other hand to create the identical pattern on the other person's arm, the one you *can* see. Weirdly, you feel that the second person's arm belongs to your body. Your creative brain, seeking to make a coherent story out of the different sensations it is receiving, assumes that the stimulus you feel and the stimulus you see must 'go together'. They must belong to the same arm. As vision is more persuasive than touch, the arm you can see becomes the one that you 'own'. Because this requires a radical suspension of disbelief on the part of your conscious self, the illusion is brittle, and can be broken if there is additional evidence that the brain's neat hypothesis can't be true. If the owner of the visible arm suddenly moves his fingers in an unpredictable way – without any corollary stimulation or motor commands occurring in your own body – the visible arm stops feeling like yours.[9]

If people's body shape or appearance changes, it's not just the 'body maps' that adapt; their apparent personality can change too, and they even think differently. When people have cosmetic surgery, or even just a makeover, their confidence often changes, and along with that their whole social demeanour. They may become more outgoing, more funny, more willing to take a risk. Just a change of clothing – those ubiquitous tools for broadcasting our values and social affiliations – can do it. Studies have shown that people wearing black clothes behave more aggressively than they do when they are wearing white. People dressed in nurses' uniforms will refuse to deliver electric shocks to other people when instructed to do so by an experimenter. The same people dressed up as policemen are more likely to agree to do what they are told. Even in virtual reality situations, monkeying about with the appearance of someone's avatar can influence how they think and behave. Nick Yee and Jeremy Bailenson at Stanford have shown that people whose VR avatars are shorter than they are in real life behave less assertively in negotiating situations, and settle for much worse outcomes, than those whose avatars are heightened. People assigned more attractive avatars become more sociable and extroverted.[10]

Distributed cognition

The backbone of the story of human evolution has been the story of perfecting our knack for incorporating an increasingly sophisticated assortment of physical tools into our increasingly flexible body schemas.

Sandra and Matthew Blakeslee[11]

We are designed by evolution to augment our on-board, physiological intelligence with all kinds of artefacts. We humans come into the world bundled with a set of capabilities – reflexes, simple skills and perceptual sensibilities. These all have their limits: I can only lift certain weights, only run at a certain speed, only hear over certain distances. So it is in our interest not just to protect these abilities but, where possible, to enhance them. Some I can develop through practice. We all learn that it is much more efficient to get about by walking and running than by crawling on all fours. Dancers and footballers develop mobility to an extraordinary degree. Masters of wine have developed their ability to distinguish subtle differences in taste. But we also have another way of augmenting our capabilities: through the use of tools. The macaque augmented his ability to reach by using the rake. I use a hammer to augment my ability to hit, and scissors to improve my ability to tear. I have spectacles that redress the decline in my on-board visual apparatus, and a mobile phone that vastly increases my ability to make myself heard over long distances. My car is a magical device that enables me to parlay my ability to fill a tank with fluid into a massive amplification of my ability to go places.

The British engineer Francis Evans has suggested that our technological disposition developed out of two evolutionary shifts. The first was an inquisitive inclination to mess about with the material we find, and get it to reveal additional affordances. If you fiddle with a stick you might reveal its ability to become a fishing rod. If you try smashing stones of a certain kind together, you might splinter off shards that can be found useful for cutting. And the second big contributor to our technological bent was, of course, learning how to get up on our hind legs and free our hands for manipulation and investigation.

Free hands vastly amplify our ability, first, to be opportunistic thing-*users*, and, second, to develop the skills to become thing-*crafters*. Increasingly adept and ingenious manipulators of material, we began to be able to craft – to an extent unmatched even by the great apes – our own worlds, and to populate them with labour-saving, intelligence-expanding devices of a thousand kinds. Find a fallen branch and it can help you climb a steep hill. Find a straighter one, with a V-shaped notch at one end, trim it a bit, and you have a much better walking stick. Find a longer, whippier one and you can learn how to vault streams with it. Invent fibreglass, learn how to mould it into a strong, light, hollow pole, and you can use it to vault over a six-metre-high bar and win the Olympics. Evans argues that we are built to be cunning exploiters of material, and thus to be able to bootstrap our own natural capabilities, physical and mental, manyfold. Prototypically, Evans recalls:

> The other day I was standing in a muddy ditch at Wortley Top Forge, and I wanted to clean earth off a stone. I glanced around and found a root – straight and strong enough to scrape with. My mind had abstracted [needed] qualities – straightness and hardness – which were unrelated to 'root', the part of a tree that sits under the ground. This mental act took place without words – readers will know what it feels like to look round the garden shed for a piece of scrap material that will 'do the job'.[12]

This is, if you will, another manifestation of our somatic nature – the deep disposition for us humans to weave together our sense of what we want to do, what we are capable of doing, and what the circumstances allow us to do. As we look around the shed in a way that is simultaneously purposeful, open-minded and wordless, our perception is saturated with those active senses of Want To and Can Do. Intelligence manifests in sophisticated seeing and ingenious doing.

Person plus[13]

We shape our tools, and then our tools shape us.

John Culkin

It's not simply that we use tools intelligently to augment our intelligence. It is more accurate to say: intelligence is an accomplishment that relies on the astute orchestration of internal loops (like the loop from the gut to the anterior insula) *and* loops and processes that connect outwards into the material world. Put simply: 'I' am only as smart as 'I' am because I am enmeshed in a vast web of smart materials: books, spectacles, notes, printers, weblinks, diaries, calendars, maps, satellite navigation screens, computer programs, filing systems, Skype links, mobile telephones . . . all of which I know, more or less, how to capitalise on. We humans are heirs to a massive cultural repository of these smart objects and instruments, and much of our learning as we grow up, both in school and out, is mastering when and how to make the most of them.

In having access to this vast backlist of useful cultural tools, we are very different from animals. But we are not completely different. All creatures are best viewed ecologically, as finding and creating eco-niches that then influence their capabilities. Here are some examples of extended systems, in which the intelligence of the whole system includes the environment as well as the animal.

A hermit crab finds a shell and inhabits it till it outgrows it, whereupon it moves, for a moment shell-less and vulnerable, into a bigger one. There is no real difference in kind between *crab*+shell, where the shell is found and appropriated, and *turtle*+*shell*, where the shell is home-grown – is there?

A spider spins its web, and then *spider*+*web* acts as the smart apparatus for feeding the spider. The web actually consists of the spider's own secretions – it is made out of spider-body – and spider and web work seamlessly together. So are there two systems at work here, or just the one?

A beaver builds a dam, which then forms a pond, which provides the habitat that is favourable for beaver life. The beaver crafts the environment, and then the environment crafts the beaver. Richard Dawkins in *The Extended Phenotype* makes a very good case that *beaver+dam+pond* work together so closely that the dam and the pond actually are parts of the extended body of the beaver.[14]

A tuna by itself is physically about seven times too weak to perform the aquabatic feats that it routinely does – accelerating like a rocket, turning on a sixpence, and so on. The way it does them is by capitalising on natural eddies and vortices in the water, and by using its tail to create additional currents which it then ingeniously exploits. The tuna uses the natural properties of the water to effectively 'turbo-charge' its own motion. Is it the tuna solo that is the 'intelligent system' here, or is it *tuna+vortices-and-pressure-gradients*?[15]

Now compare these extended eco-beasts with a hypothetical human Alzheimer's sufferer, already with significant memory loss. Many such people manage to maintain a high level of functioning within the normal community by deploying a range of external props and aids to help them. These may include labelling the objects around them, using a 'memory book' with annotated photographs of family and friends, or a diary for routine tasks and events, and adopting simple tactics like leaving things they are likely to need in plain view, so they will be easy to find when the occasion arises. They are cleverly using the environment to offset the increasing limitations and fallibilities of their own biological system. Is looking up an address in your own notebook so different from looking it up in your on-board memory banks? Is *patient+mnemonic-devices* any different from *tuna+currents,* in the way they go about being intelligent?

Now a more hi-tech example. Imagine you are strolling down the main street of your home town trying out your new third-generation competitor to Google Glass. Let's call them Mnemoptics. They are connected wirelessly to a cloud facility that contains face recognition and navigation software, and a personal database of images of your friends and acquaintances, together with biographical information, such as the names of their spouses and children, the occasion on which you last met, their favourite food, and so on – which you have uploaded and keep updated. As you walk, the software scans the faces in the crowd, and when it finds one it recognises, it displays their fact sheet on your 'Autocue' lenses. Social embarrassment is a thing of the past, and old acquaintances ought to be hugely impressed by your recall of their love of seafood and the name of their latest grandchild – except that they are wearing Mnemoptics too. The fact that your enhanced memory loops out into the extra-corporeal world, rather than remaining confined to the biological body-brain, seems hardly to matter. It's not really different from the attempt to improve your memory by using 'cognitive enhancing' drugs, say, or by taking daily exercise to keep those little grey cells in good condition – is it? Are we not all 'intelligent' these days because we are *person+hand-held-device*?

Finally, a very lo-tech example. If I want to know the answer to '8 × 7', my brain pops the answer into consciousness without any fuss or delay. I've learned it, and the calculation is recorded in my circuitry. But the answer to '362 × 89' is not. I have to work it out, and to help me do it (the old-fashioned way), I use a piece of paper and a ballpoint pen to record and accumulate the results of a number of component calculations to which I

do have ready-made answers. Unable to hold all these components in mind as I go along, I use the paper and pen to offload
the memory demands. The 'cognitive process' seamlessly inter-
weaves internal and external processes; I am thinking, remembering and interacting with the pen and paper all at once.
Though philosophers are haggling about the niceties, it seems
perfectly reasonable to me to see this orchestration of thought
and action, memory and writing, brain and paper as essentially
Mind At Work. We routinely sidestep the shortcomings of our
own on-board intelligence by intelligently co-opting bits of the
world, and intelligently using them to amplify our capabilities.
We are constantly on the lookout for the next mindware
upgrade (as Andy Clark puts it) to come along.[16]

Mundane though this last example is, it highlights one of
the most important external amplifiers of human intelligence:
writing, and the ability to make all kinds of stable records of our
mind's 'work in progress'. Speech had given our ancestors
enhanced abilities to coordinate actions and share information.
But the discovery and invention of tools that make marks,
whether as sketches on a cave wall or hieroglyphics on papyrus,
was momentous for the expansion of our intelligence. It became
possible routinely to offload memory demands: to write shopping lists, for instance. It is not only Alzheimer's patients who
benefit from such mnemonic prostheses.

But more importantly, we became able to reflect on our own
developing ideas. Tentative formulations could be recorded and
returned to with a fresh mind or a new perspective. Making
sketches, models and drafts enables us to 'freeze-frame' our
intelligence-in-action. Writing and drawing let us take time to
reflect on our 'work-in-progress', in a way that on-line, real-
time thinking cannot, and thus come up with better products
and solutions. They vastly expand our power of being able to

talk to ourselves. And, of course, writing and sketching make evolving ideas available to other people, and thus enable a wide – now worldwide – circle of critics, sounding-boards and collaborators to contribute to the development of those ideas. Learning to create lasting embodiments of fleeting thoughts made available whole new ways of being cumulatively, cooperatively intelligent.[17]

These examples just point up what we all do all the time. We, like the beaver, construct our worlds to be full of performance enhancers. Some of them interface with the body very directly, like heart pacemakers, cochlear implants and prosthetic limbs. Others are at one remove, like smartphones, kitchen blenders and hockey sticks. Some tools help us do practical things – cook sous-vide steaks, record how far and fast we have run. Some enhance our perception and communication – semaphore, television, loudspeakers; while others help us learn and think. Amongst the latter are sketches on the backs of envelopes, scribbles in the margins, calculators, filing cabinets, libraries, the internet – and the semi-organised piles of notes and articles laid out around my chair on my study floor. If someone can tell me where to draw the line between aided and unaided intelligence – between person-solo and person-plus – I'd be glad to know.

The fact that our tools 'become us' also begins to explain why people grow so fond of and dependent on them. Chefs become attached to their knives, musicians to their instruments, artists to their brushes, carpenters to their chisels and millions of motorists to their cars. Tools become, psychologically as well as neurally, so much a part of us that their loss feels like an amputation. There is, these days, almost no difference in kind between losing your smartphone and having a mini-stroke. If someone were cruel enough to steal a blind person's white stick, or to deliberately remove all the clever devices that the Alzheimer's sufferer has painstakingly crafted to help her function, it would not be theft, a crime against property, but violence, a crime against the person.[18]

Social resonance

If I wish to find out how wise, or how stupid, or how good or how wicked anyone is, or what his thoughts are at the moment, I would fashion the expression of my face, as accurately as possible, in accordance with the expression of his, and then wait to see what thoughts or sentiments arise in my mind or heart, as if to match or correspond with the expression.

Edgar Allan Poe

We have just looked at a loop that couples my mind with a piece of paper and some arithmetical calculations. I think and write down some numbers; and then I read the numbers and they feed back into my calculating ... The paper and I are in dialogue; we reciprocate. But that loop looks really slow and clunky compared to the speed and intricacy of the resonance that connects me dynamically with another human being. If we are built to amplify ourselves by turning objects into tools and resources, how much more are we designed to reverberate with other people.

This happens quite automatically at the neurochemical level. Remember, our brains are peppered with 'canonical' neurons that automatically fire up actions relevant to a seen object, and 'mirror' neurons that fire when we do something, and also when we see someone else do the same thing. So, when I see you pick up a cup, my brain automatically primes me to copy you, and when I see you smile, I am already halfway to smiling back. (Sometimes my brain primes me to perform an action complementary to yours, rather than the exact same one.) Vittorio Gallese, one of the original discoverers of mirror neurons, has argued strongly that we are naturally inclined to look at what people around us are doing, and automatically convert these perceptions into a variety of internal echoes. We are inclined to mimic their gestures and facial expressions, for example (or to reciprocate, as in the childhood game of Peek-a-Boo).

Through this internal resonance, our system also recruits the visceral motives and concerns that we habitually associate with those gestures and perceptions. And we then use these as a basis for understanding what the people around us are up to, and how they are feeling – exactly as Edgar Allan Poe indicated, except that he offers it to us as advice, which we can follow or not, while recent evidence shows that imitation is our 'default mode'. It is only with the gradual development of inhibitory control by the frontal lobes that we learn to restrain and restrict the mimetic impulse.[19] When people lose the capacity for inhibition through damage or disease of the frontal lobes, they become compulsive mimics of other people's gestures and speech. In a disorder called echolalia, their imitation may even be triggered by their own speech or behaviour, locking them into long, debilitating loops of self-imitation.

So whether we are aware of it or not, our bodies are in a state of continual resonance with those around us – or those we may be remembering or imagining. While you and I think we are discussing the film we've just seen, our bodies are dancing with each other's every gesture and expression. If we are sitting side by side in rocking chairs, and able to see each other, the rocking of our two chairs will synchronise, without our awareness or intention, even if one chair is weighted so that it requires more effort to rock it. In fact, as we were watching the movie, our brains quickly developed a point-by-point synchronisation. We were literally entrained.[20]

Good communication depends on this bodily coupling. A study by a research group at Princeton put pairs of people in linked MRI neuroimaging machines, and had one of them relate an unrehearsed, real-life story as if speaking to a friend, to the other person. (In one example the narrator was an under-graduate telling an embarrassing story about her high-school prom.) They found that the activity in the listener's brain mirrored the speaker's brain activity, usually with a delay of a second or so. In the neuroimagery records you can actually see the listener reconstructing a model of the narrator's story in his own brain, as they go along. However, the researchers found some areas of the brain where the synchronised activity in the listener's brain actually preceded the corresponding utterance by the speaker. As we might expect, the higher levels of the listener's brain, located in the prefrontal cortex, are trying to anticipate what is coming next and, if he gets it more or less right, his provisional, predicted story structure matches what the speaker actually says. These brain areas are the same ones that are involved in inferring other people's beliefs and motives, and it is these that provide the main scaffolding around which the developing story is constructed. The stronger the match between what goes on in the two brains, the better is the listen-er's understanding. If the two brains fail to synchronise, communication breaks down.[21]

And it's not just brains that need to synchronise if commu-nication is to succeed; bodies do, too. Swiss researcher Fabian Ramseyer analysed videotapes of a large number of psycho-therapy sessions and found that the extent to which client and therapist mirror each other's body language predicts the client's satisfaction with the therapist, and the strength of the bond that the two of them have formed. This mirroring included coordinated body movements and gestures, the congruence of their posture, mimicry of each other's facial expressions and the adoption of similar voice quality.[22]

When we say (in a parody of the empathetic Californian) 'I feel your pain', we often mean we are resonating at the emotional level. But a recent British study shows that, when we are watching someone else in pain, many of us feel comparable sensations in our own bodies. In this study, people were shown either film clips or still photos of other people in pain, and about a third of them reported physical sensations – always in the appropriate part of the body. The intensity with which we react to the sight (and sound) of someone else in pain depends on how emotionally close we are to them. If you watch a nurse stick a needle into your own child, your involuntary reaction is (not surprisingly) much stronger than if it is someone else's child. We resonate most strongly with those we care most about.[23]

Most of this social resonance slips by without our being aware of it. If you flash an image of fearful eyes to people so quickly that they are unaware of them, their brains nevertheless respond (see Figure 11).[24] If you vary the size of the pupils in a variety of photos of people's faces, an observer judges the mood and character of the people differently, yet they have no awareness of the influence that the pupil size is having on their reactions. And the observer's pupils tend to dilate or contract to match the person they are looking at – again without their conscious awareness. If the picture of a woman on the cover of a psychology textbook is altered so that her pupils are enlarged slightly, male undergraduates buy more copies of the doctored version.[25]

In a more real-life test, Eunhui Lie and Nora Newcombe at Temple University showed nine- and ten-year-olds photos of

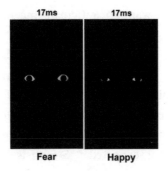

Fig. 11 Scared eyes will activate the amygdala, even subliminally.

children who had been at kindergarten with them, mixed up with some similar photos from a different kindergarten. Very few of the older children consciously recognised their former preschool-mates – yet their skin conductance showed a very clear hike when they looked at the faces of their old peers but not when they looked at the others. All the time, it appears, we are sensitive and responsive to those around us in ways that our conscious mind simply does not notice. We are hooked up to each other in the way that cell phones are: we are in touch, silently (and sometimes audibly) vibrating in response to 'calls' that we may or may not choose, more consciously, to answer.

This instantaneous resonance doesn't just make us feel close and connected, though it does do that. It enables us to coordinate our actions more successfully with other people, because we are better able to predict what they might be about to do, and therefore how they are likely to react to what we do. I can assist you better if I can get inside your skin and feel your state and your intentions. People who know each other well are notorious for finishing each other's sentences. And, if our relationship is of a different kind, I can outwit you better. I can use

213

your tendency to want to anticipate me to trip you up by encouraging you to see me one way – and then doing something completely different. Sun Tzu's legendary sixth-century BC book on *The Art of War* counsels the military leader to 'know your enemies, and know yourself . . . If your opponent is temperamental, seek to irritate him. Pretend to be weak, that he may grow arrogant. Attack him where he is unprepared; appear where you are not expected.'[26] Scientists such as Nicholas Humphrey have even argued that it was this kind of 'arms race' of social intelligence, driving ever subtler attempts at anticipatory cooperation and competition, which accounted for the sudden evolutionary growth-spurt of the human brain.[27]

The origins of resonance

There are no chaste minds. Minds copulate wherever they meet.

Eric Hoffer

People's intricate ability to predict and attune with each other has its origins, of course, in the interactions of babies and their mothers. Even before birth, the baby's body-brain is tuning itself to the rhythms and habits of his mother. After he is born he will prefer songs his mother sang while he was in the womb to other songs. In her arms, he will learn to adjust himself to her smell, breathing, postures and movements. He will soon come to know the rituals of feeding time, and will ready himself in anticipation. It is as if mother and baby are dance partners, learning each other's moves, so that, if all goes well, they do not tread on each other's toes. They become familiar with each other's natural dance steps, and accommodate accordingly.

Each mother–baby pair develops its own dance, but, provided they are sensitive to each other, the precise form of their dance is less important than the fact that they have one. A case study of a sighted baby with two blind parents, for example, found that the form of their social dance was obviously different from usual – they learned to waltz, so to speak, while most babies are learning a version of the quickstep – but the child's development was not hampered by this difference at all.[28] Children of mothers suffering from postnatal depression, on the other hand, often do show a delay in their social, emotional and mental development, and this is because the mother, preoccupied with her own troubles, may be less sensitive to her baby's signals, and thus the dance – any dance – fails to develop.[29] In general, one of the effects of depression, in both children and adults, is to dampen this social resonance so that the sufferer actually feels physically isolated from others (and they from her).[30]

As we grow up, this innate propensity for social resonance becomes customised: we become attuned to different individuals, and resonate differently in the light of our past experience with them. Through repeated interactions we discern their traits and habits and build these into a neural model, within our body-brains, that enables us to predict how those individuals will behave in a whole variety of circumstances. Just as the abstraction 'Timmy' (which we met in Chapter 7) became the web of expectations relating to next-door's cat, so 'Mummy', 'Daddy' and 'Nanny' become models for guiding actions with significant others. As predictions are proved accurate or inaccurate in the light of events, these models become more adaptable and reliable. And as we meet a widening range of others – classmates, their parents and siblings, teachers, as well as heroes and villains in literature and film – so these models become more numerous. In deliberate, conscious mode, we may be aware of asking ourselves, especially in a tricky situation, 'How would Aunty Helen have dealt with that?' or even, perhaps, 'What would Jesus do?' But long before we are capable of being so explicit, our body-brains have been busy building on our inherent ability to do the same thing implicitly.

Because our concepts interweave perceptions, actions and concerns (or Needs, Deeds and See'ds, as I put it earlier), these models of familiar individuals enable us to predict not just what they will do, on certain occasions, but how they will see things, and how they will feel. So these models can underpin the child's growing powers of empathy – having an increasingly good idea what it is like to be 'you' (and 'you' and 'you' and 'you'), and thus to adopt different perspectives on events. I become increasingly able to detach myself from my default, egocentric constellation of habits and concerns, and see the world through other people's eyes. A child becomes better at hiding as she grows in her ability to adopt the perspective of the 'hunter'; better at comforting others as she realises that not everyone shares her own portfolio of anxieties and reassurances. This ability to adopt other perspectives expands her social intelligence; if she and I can see the world from different perspectives, we are less likely to be locked into our own, and therefore better able to find common ground on which to meet. And more generally, intelligence is expanded by the ability to think about and imagine situations from different points of view.[31] As British neuroscientist Chris Frith puts it, rather more formally:

> Of all the representations held in the brain, that which is coded in non-egocentric coordinates will most closely resemble that held in the brain of another. It is these representations that will best enable prediction of the behaviour of another.[32]

As the sophistication of my mental models grows, I become able to incorporate your model of Me inside my model of You. I can begin to imagine how you see me, and what you think of me. When I inhabit your vantage point, one of the things it enables me to see is myself. I can become an 'object' to myself, with traits, temperament and habits as well as a visual appearance. (Mirrors enable me to see a version of the face that you see when you look at me. I have to paint in the expression of excitement and anticipation, tinged with a touch of anxiety, that I always wear when we meet.)

We could, without too much fancy, imagine that these conceptualisations of individuals could give rise to more general images of *kinds* of others, based on higher-level abstractions: people who give me a hard time and make me feel guilty; people who seem sympathetic and understanding; people who frighten and confuse me; people who will rescue and comfort me; people who can offer wise advice when I have a problem; people who are charming but untrustworthy; and so on. These high-level categories of significant others begin to look rather like Jung's 'archetypes', and several authors have begun to see if these archetypal, mythic characters – the Judge, the Wise Old Soul, the Witch, the Saviour, the Trickster – can be given bodily underpinnings.[33]

I would like to pick out one of these putative archetypes for special attention as it will be useful to us later. I call it the Benign Generalised Other: the image of a person who knows us deeply and judges us not at all. Sometimes this image derives from a grandparent, or from a counsellor, psychotherapist or priest. Sometimes it is embodied in a religious figure such as Christ or Buddha. Often it is a composite of several sources. Whatever its provenance, it has the benefit, if one can learn the trick of mentally 'putting oneself in their shoes', of providing a warm, neutral point of observation for our own and others' behaviour.

This attentive, accepting vantage point lies at the heart of what is fashionably called 'mindfulness' practice.

*****

In terms of their substance, our bodies look like disconnected things. But functionally, in terms of our process, we are distributed beyond the contours of our skin. Each moment our bodies are sensitive to influences from outside which are constantly conditioning and nudging what goes on within. Through the body we and the world are intimately and dynamically interconnected. And this affects not just what we do, and how we relate, but the products and performances which we generate. In craftsmanship, as we work material we can feel this connectedness quite intensely. That's what we'll look at next.

10

CRAFTINESS AND EXPERTISE

That it can be as much a central mark of intelligence to be able to improvise a useful or decorative artefact from available materials, as it is to be able to talk about the problem to others, has not generally been seen by philosophers to invite any special sort of enquiry.

Andrew Harrison[1]

Our core design specification, developed through hundreds of thousands of years of evolution, is to find opportunities to do things that matter. All our sub-systems are yoked together by the overriding concerns and interests of the body as a whole, and of the social and biological niches in which we are embedded. Deep down we are doers and makers, practical problem-solvers, not thinkers and talkers.

Thinking and talking have evolved as powerful adjuncts to these embodied drives and capacities. They do not stand alone, nor do they supersede the requirements and accomplishments of the body. If we do not see our intellectual functions in this light, we see them wrongly. Our bodies are composed of many muscles and organs that have precious little to do, directly, with reasoning and understanding – our calf muscles and solar plexus play little obvious part in the solving of a crossword puzzle. But the fact that we are composed of all these functions and capacities profoundly influences not just what we *can* do, but what we might *want* to do. A renewed appreciation of the centrality of the body colours the way we think about how people spend their lives: the skills we invest in developing; the projects from which we derive satisfaction and fulfilment. Without question, some of these are intellectual. People grow the skills of public speaking, or of writing, and derive great pleasure from reading and debating. In the digital world, people invest huge amounts of time in mastering complex computer games and in browsing the Web.

But it is also true that much voluntary energy goes into developing physical prowess and creating concrete achievements. In a small unscientific survey, I have unearthed university vice-chancellors whose passion is for windsurfing; senior civil servants who live for their rose gardens; successful writers who derive as much satisfaction from producing a five-course meal as from having a book nominated for a prize; educational psychologists who are mad about horse-riding; chief executives (male) whose cupboards are full of home-made jams; 50-year-olds whose weekends are still mostly about music, dancing and festivals; television presenters who spend all their free time fishing, or who hide away in Devon making beautiful bowls with old-fashioned, handmade pole lathes. I have a friend with a high-up job in the art world in London, her weekdays full of important meetings and position-papers to trustees, who labours for eight hours in her garden on Saturday, in the rain if need be, and comes in exhausted, muddy and with a deep glow of happiness on her face. I still have a bedside table and a small chest of drawers that

I made at school sixty years ago, and of which I am unaccountably proud. Political economist Francis Fukuyama made his family's dining table and his children's beds. 'Few things I've created have given me nearly as much pleasure as those tangible objects that were hard to fabricate and useful to other people,' he says.[2] There is something about our skilful, muscular involvement with real material that will not go away – because we are deeply configured for it.

Intelligence in the making

It is easy to see, and appreciate, the skill and sensibility of a violin-maker or a sculptor, or the grace and virtuosity of a gymnast or a pianist – but what about the somatic intelligence of the manual worker? If we really are built to make and do, we have to see these qualities in the everyday as well as in elite forms of accomplishment. Take, as a case in point, Rose Rose. Rose was a waitress in Los Angeles for much of her life. Born Rose Meraglio, she met Tommy Rose in the 1940s and had a son, Mike, who is now a researcher at the Graduate School of Education at UCLA. Through watching his mother at work, Mike became intrigued by the unsung intelligence that she displayed. This interest grew into a broader research project, and later a book, *The Mind at Work*, in which an ethnographic study of his mother constitutes the first chapter. Any business traveller who regularly dines alone may well have made similar observations to those recorded in detail in the book.[3]

When restaurants are busy, the servers work hard and many of them are very subtle and skilful in what they do. First of all there are the physical skills they have learned. Rose Rose was small but she could carry seven plates on her right arm and two cups and saucers of coffee in her left hand, moving at full tilt through a crowded space, while not spilling a drop of coffee or getting food on the bottom of any plate. That takes degrees of strength, balance, fine motor control and all-round perception that bear comparison with those of a gymnast or a dancer. You have to lock your arm, torso and back in a particular fashion while walking in a way that minimises the build-up of fatigue. Rose could be on her feet for hours on end. Another waitress-researcher, Lin Rolens, explained: 'You learn a walk that gets you places quickly without looking like you are running ... This requires developing a walk that is all business from the waist down, but looks fairly relaxed from the waist up.'[4] These skills were consciously practised by Rose in her early years of waitressing.

Then there is memory. A good server like Rose can relay complex orders from large tables to the kitchen without the aid of a notebook and, when serving, remembers who wanted the burger without lettuce and who asked for two straws with the Coke. Studies of servers show that they employ an intricate set of mnemonic strategies that drastically expand the normal limits of 'working memory'. Rose would use private imagery of people's distinguishing features, their clothes and so on, and attach an image of their order – sometimes in funny or irreverent ways. She could create a narrative of where different people were sitting round the table, drawing on physical location and, for example, who it was who had drawn up the fifth chair at a (normally) four-seat table. If it's the woman who ordered the steak and beer, and the man the salad and wine, she would note and make use of that. She would quickly bring back the little container of maple syrup and place it so that it identified the person who had ordered the waffles. If someone was pleasant or unpleasant to her, she would use that too to help the order to stick.

Social intelligence is paramount for Rose; she wants to create a pleasant experience for her customers, not least because her income depends on tips. She has become skilled at being pleasant and personal, attentive to individuals and not trotting out tired phrases ('Everything all right with your meal?', 'Need any sauces at all?') while not overstaying her welcome and remaining efficient. She would spot, out of the corner of her eye, the customer whose order is delayed, check his body language, and plan her next route through the restaurant to enable her to pass by his table and have a quick, sympathetic word. Every trip round the restaurant she makes is done with an eye on harmony and efficiency. A colleague explained, 'I think of it as a kind of ... flow of organising what can be done in one full circle [round the restaurant]; how many tasks can be accomplished, as opposed to back and forth, back and forth ...'

In general, life in the restaurant is complex, and sophisticated planning essential. Rose describes a typical situation where 'an obnoxious regular is tapping the side of his coffee cup with a spoon while she is taking an order at an adjacent table; the cook rings her bell to indicate that another order is ready; the manager has just seated two new parties at two of her tables that have just cleared; and, en route to the table that is now ordering, one customer has stopped her to modify his order, another signalled for more coffee, and a third requested a new fork to replace the one he dropped on the floor.' She comments, 'Your mind is going so fast, thinking what to do first, where to go first, which is the best thing to do, which is the quickest ...' Another waitress commented: 'As you walk, every time you cross the restaurant, you're never doing just a single task. You're always looking at the big picture and picking up things along the way.' And this, maybe, for several hours without a break.

This complexity is typical of intelligent functioning in real situations. The scholar can usually sit quietly in her study and concentrate more or less exclusively on the PowerPoint presentation she is creating for next week's lecture. A person taking an IQ test or sitting an exam is not simultaneously being bombarded by the invigilator with emotionally tricky demands. But this seclusion is in many ways unlike the real-world hurly-burly within which we are mostly required to act. Like Rose, we are often juggling calls on our attention, physical demands, social interactions and emotional and visceral promptings all at once, sometimes under severe time pressure, and with real costs and consequences if we get it wrong. Conscious thinking and reasoning are part of what Rose does, but they constitute only one section of the psychological orchestra that is frequently playing *tutti* and *fortissimo*. Her intelligence requires full-on coordination of mind and body, action and reflection, feeling and cognition.

In another chapter in *The Mind at Work*, Mike Rose uses a more traditional exemplar of the manual crafts and trade – carpentry – to take a close look at the kinds of learning that are needed to build up both the expertise and the identity of a 'craftsman'. Under the careful eye of Jerry Devries, a group of apprentices are learning a great many things in ways that involve as much body as mind. They are learning an appreciation of material: what different kinds of wood are good for, and how to work with different grains and knots. They are finding out what oak does easily, what it can be pushed to do, and what it is unwise to attempt. They are developing sensitivity, getting a 'feel' in their fingers for how wood responds to the tools they are using: how a chisel might easily jam in walnut or skid in sycamore. They are learning to see when a joint is tight and an edge is planed straight. They are learning, sometimes the hard way, the importance of the old carpenter's adage, 'Measure twice, cut once'.

They are learning to balance strength and gentleness, speed and patience: to hear when a drill is being pushed close to its limit; to know how much force will cause the bandsaw blade to twist. They are learning how to rectify different kinds of mistakes, and what to do when they hit what Robert Pirsig in his classic *Zen and the Art of Motorcycle Maintenance* called 'gumption traps': botches, usually of your own making, that make you feel stupid and angry, and likely, if you are not careful, to behave even more stupidly and make matters even worse. They are learning to recognise when they need to ask for help. (In a small piece of research we carried out at the Centre for Real-World Learning, we asked a group of plumbers what was their most frequently used tool. The answer: their cell phone.) And the apprentices, in their groups, are learning how to interweave their youthful banter with a serious-minded give and take of information about the job at hand.

They are learning what different tools do and what, at a pinch, they can be made to do. They are learning how to care for tools, the rituals of oiling and sharpening, and the importance of doing this. They are allowing their body-brains to get used to the behaviour of different tools so that, when they are picked up, they immediately become a fluid, functioning part of the body (as we saw in the previous chapter). They are learning to account for the width of the saw-blade in their measurements and calculations. And they are learning to plan and think strategically. Rose quotes one old-timer, saying, 'You develop a sense of anticipation. The idea is always to think ahead, visualise where you'll end up, what you'll need next, and next after that.' The development of visualisation or mental rehearsal as a vital 'tool' in its own right, in the craftsman's psychological repertoire, is acknowledged in many forms of physical expertise.[5]

We should also note that craftable material comes in many forms: clay, food, words, wood. But the material closest to hand is, of course, our own bodies. For many people – not only the young – their most intricate, absorbing, extended 'work in progress' is their own appearance. Through tattoos, piercings, cosmetic surgery, diet and fashion, one can work the material of one's own body into creative forms of self-expression and identity. People become their own most highly crafted product and performance. Nothing wrong with that, but it becomes so prevalent and so obsessive for some, perhaps, precisely because the opportunities for physical making and doing in schools and elsewhere are held in such low esteem.

Working knowledge

Builders and carpenters need to learn a lot of mathematics. They need knowledge of geometry, measurement, ratio and proportion as well as arithmetic. But they need to be able to make use of this knowledge in the context of rich, complicated problem-solving, under pressure, often with the participation of others. The maths is interwoven with considerations of available time, cost, conversation and aesthetics. Tried and tested rules of thumb happily rub shoulders with more abstract systems of symbols, and the latter are not universally superior to the former. The Cambridge mathematical genius Alan Turing took two months to figure out why his bicycle chain fell off when a rare conjunction of spokes, cogs and chain links occurred: something that a competent bicycle mechanic would have sussed, using far less powerful but much more useful heuristics, in two minutes. In a similar vein, I once went to a pub in Fulham with an American Professor of Mathematics Education, where we played darts. We were well entertained by how bad he was at scoring. Locals could calculate 501 minus (3 × 18) minus 5 in the blink of an eye, while the prof was reduced to searching for pen and paper.

The irrational denigration of 'craft knowledge' and 'know-how' runs back to the ancient Greeks. Mike Rose remarks:

> In *The Republic*, Plato mocks the craftsman who would pursue philosophy, for his soul is 'warped and maimed' by his work; such men are 'incapable of culture'. And Aristotle in *Politics* notes that 'there is no element of virtue in any of the occupations in which the multitude of artisans and market-people, and the wage-earning class, take part. Because such occupations are 'ignoble and inimical to goodness', Aristotle further proposes that their practitioners be denied citizenship . . . Plutarch wrote, 'It does not necessarily follow that if a work is delightful because of its gracefulness, the man who made it is worthy of any serious regard'.[6]

From the Greeks we have inherited the iniquitous tendency to assume that, wherever mathematics, for example, is made practical and applicable to concrete situations, it should be credited with less prestige, and the user of this practical knowledge is supposed to need less intelligence. There is also a strong tendency to assume that intellectual understanding is sufficient, and often necessary, for the proper deployment of knowledge in action. In the Platonic–Cartesian world, comprehension needs to precede, and is generally senior to, competence. In school mathematics, this view led to the assumption that children somehow needed to have 'understood the concept of number' before they could reliably do their sums. That this abstract endeavour might be merely interesting to some children, rather than a necessity for all, seemed not to occur to enthusiastic proponents.

227

From an embodied point of view, however, the ability to articulate what you are doing, and why you are doing it, may or may not be necessary or useful. Many virtuoso musicians (in many different genres and cultures) are unable to read music or discuss composition in learned terms. Many prize-winning gardeners are not handicapped by ignorance of the chemical equation for photosynthesis. Many top athletes and sports-people understand the body, its muscular tolerances and nutritional needs, in a piecemeal fashion that is entirely adequate. But when these experts become coaches and mentors they may well find that a richer vocabulary and a deeper, more coherent understanding really are necessary. The amount and nature of the explicit knowledge you need is entirely dependent on the job; it is not a universal good.

*****

Jeanne Bamberger, a professor of education at MIT, used to run a 'Laboratory for Making Things' for school students aged from 6 to 14 in Cambridge, MA. Some of the challenges she offered to the children who came were to make various kinds of mobiles, or to design and fit a gate. She found that some children were highly articulate about the laws of balancing – but were hopeless at translating this knowledge into making mobiles that actually balanced or gates that swung properly. Many of them were certified as being 'gifted', but when it came to pragmatic action they were just the opposite. Practically, they were 'extremely challenged', had 'severe learning difficulties', and 'exhibited special educational needs'.

And there were other children who were the reverse. They often had a sophisticated 'feel for' the mechanics of what they were trying to achieve, but found it very hard to translate that 'feel' into words. Ruth had been categorised as having serious difficulties with language, both speaking and reading, but the gates she made were really intricate and clever. When asked to explain what she was thinking, she would mumble and quickly give up and say things like 'I don't know' or 'Oh, just forget about it.' But with a good teacher, and a patient group of other children, she began to talk about what was in her mind.

It turned out that Ruth had developed, to a very high level, the engineer's and architect's ability to 'mentally feel' the properties and predict the behaviour of the structures she was making. Often she couldn't find the words, though her gestures would give much stronger clues to the complex system of relationships she had in her head. With encouragement, she grew in her ability to 'unfurl' her bodily understanding into words – and to discover two things: first, the value of being able to bring the manual and the intellectual closer together; and second, the value of being able to learn with and from others. In one review session, after coming to the lab for a couple of years, Ruth said: 'It's so funny to compare ways 'cause you think you've got the right answer, but you don't have, you've got one answer. And when you talk about it, you learn more from yourself. 'Cause when you talk about something, when you say it out loud, you kind of think of it yourself in different ways . . . And explaining it you feel good . . . you get what I mean?'[7]

There are many ways of thinking, and thinking with your mouth or your laptop are only two of them – albeit important ones. People think with their bodily feelings. They think with their mind's eye. And they think with their hands, both through gesture and in the actual process of making things. It is not that the intellect conceives of things and then the hands do what they are told. The interaction of the two is (to revisit an earlier analogy) more like a dance in which sometimes one person leads and sometimes the other. The English artist Grayson Perry says, 'Ideas come on the hoof, while I am working with material.'

In his book *The English Patient*, novelist Michael Ondaatje coined the word 'thinkering' to express the way in which the hand feeds the mind – the world of possibility – just as much as the mind instructs the hand – the world of actuality. Artists develop their ideas through sketching; scientists through building models of chemical reactions; and writers through the process of writing, drafting and editing. As I write this book, much comes out through my fingers and on to the screen that did not previously go through my mind – and then I can see what I have written, which may not be very good, but it has the spark of an idea that I can then refine. The physical process of writing is an inextricable part of my process of thought. I have some ideas and resources when I sit down to write, but what appears on the screen is often a surprise, and sometimes serendipitous.

Both Michael Faraday and Thomas Edison, creative scientists and inventors extraordinaire, had little formal schooling – though they read voraciously. They started their careers as technical apprentices. Faraday was hired by Sir Humphry Davy as his lab technician and valet, and swiftly grew to become Davy's fellow researcher. As far as we know, Faraday never wrote an equation in his life, but that didn't stop him having the insight behind Faraday's Law. Michele and Robert Root-Bernstein (from whose stimulating blog this story is taken) skewer the main point like this:

Physical manipulation of things, like direct personal experience of any kind, generates sensory images of all sorts and thus enables thought. Hands-on tinkering leads to minds-on thinkering. Bodily engagement with nature teaches much more than any amount of words or numbers in science books. Doing produces a personal understanding that symbols simply can't.[8]

The limitations of intellect

Sometimes understanding may even interfere with expertise. We saw earlier that novice golf putters whose heads were full of useful ideas and explanations went to pieces more disastrously when they had to perform under pressure than did those who had no such knowledge 'to fall back on'.[9] Other studies have shown that people who 'have the theory' may do better in routine situations than those who don't, but when things become irregular or counter-intuitive, understanding becomes a handicap. Such pepole cling too long to a mental model that isn't working, or keep trying to make it fit the new world. People who don't have the theory are capable of being more 'present' to the new situation, more attentive, and thus they notice new patterns and extract vital information faster.[10]

Indeed, even when the theory is fitting, there comes a time in the development of expertise when it has to be transcended. American philosopher Hubert Dreyfus has identified five rough stages that someone traverses on their way from being a novice in a particular sphere to becoming an expert. They differ in the kind of thinking that has to be done. *Novices* remember and follow 'the rules'. *Advanced beginners* still have rules and maxims to follow, but the stages become more subtle and situation-dependent. You have to think: 'What kind of situation is this?' and *then* apply the rule. When you are *competent*, you don't classify situations according to type; you think for your-self about what the most important priorities are, and then apply the skill and knowledge you have already gained. When you become *proficient*, you need to think less often, as your ability to 'read' situations becomes more immediate and intuitive, but you may still need to refer back to the 'rule-book' (in your head, if not literally) to check on procedure.[11]

And finally, the *expert* may scarcely need to think or analyse the situation at all. Their experience is so rich, varied and deeply embedded that, like a top-flight tennis player, they are able simply to respond by doing the right thing intuitively – at least 99 per cent of the time. They are picking up subtle cues that they may not even be aware of, and which may be too nuanced to have found their way into any rule-book, and they may be using these to guide their embodied reaction: their answer to the perennial question, 'What's the best thing to do next?' Another influential philosopher, Michael Polanyi, famously called this *tacit knowledge*: knowledge that is such a fine web of contingent possibilities, built up through years of experience, that it simply cannot be rendered down into words. The neurochemical loops and networks that underpin your expertise are orders of magnitude more intricate than any vocabulary, however technical, could hope to capture. It is not, like Ruth, that you are inarticulate; the knowledge itself is of such delicacy that it is in principle inarticulable.[12]

This means we have to acknowledge and value ways of knowing that are not capable of being unfurled into language, because any language, like the symbols on a map, creates distinctions that are not actually present in the neurochemical territory of the body. All maps have to be crude (mis)representations, or they will not be useful for finding our way about. As experts, we do the right thing – we are intelligent – but we are at a loss to explain how or why we did so. On this view, adherence to conscious clarity and explanation holds learning back.

A teacher or an education system that insists on testing explicit understanding therefore arrests learners' development at an intermediate stage of mere competence or mediocrity. Dreyfus summed it up at a symposium in Birmingham in 2008 that I was lucky enough to attend. 'If you can explain it, you don't *really* understand it,' he said.

Nevertheless, as with Ruth, the struggle to get closer to being able to articulate your tacit knowledge, while ultimately doomed, may be very productive. It makes you think, and it connects you with other sources of support and criticism. Any deep process of verbal creativity will, of necessity, involve an uncomfortable groping for phrases and images that try to do justice to something 'real' in the territory that cuts across the conventions of the map. As e.e. cummings says in his 'Poet's Advice to Students':

Whenever you think, or you believe,
or you know, you're a lot of other people; but the moment you
feel, you're nobody-but-yourself . . .

[And] as for expressing nobody-but-yourself in words, that
means working just
a little harder than anyone who isn't a poet can possibly
imagine . . .[13]

Learning

How do we train our bodies to do so many intricate and clever things? Even walking, skipping, throwing a ball and chopping vegetables are fairly impressive achievements of coordination and control, let alone winning Wimbledon or Young Musician of the Year. The first thing to say is that embodied learning is quite unlike school learning. It's your developing ability to *do* that counts; you don't get many points, in your football team or choir, for being able to talk or write *about* what you are doing. I remember, when I was running a teacher training course many years ago, regularly witnessing the shocking realisation, for many eager new trainees, that the learning strategies that had led to their 'first' in History or 'good 2:1' in Chemistry were not going to work when it came to learning how on earth to keep a boisterous class of 13-year-olds under control. Their well-honed abilities to debate, analyse and explain were not going to get them very far. And, as learners, they did not have much of an idea about what it was that they lacked.

So let us have a quick look at how non-intellectual learning happens. There are, as we would expect, three components to expertise: having a good repertoire of reliable, flexible, skilful *actions* to call on; being perceptive about the *opportunities* for action and the *affordances* of the material you are working on; and hooking both of these up with a sense of what kinds of *valued outcomes* could arise from acting that way. Learning is developing each of these in concert with the others. Obviously, there is no point in knowing what you want if you have no means of achieving it, or in knowing what you can do but not when it is appropriate or convenient to do it. But, for the sake of description, let's pull them apart a little just for a moment.

There is *learning by noticing*. Just by being attentive to the world, the neurochemical systems pick up patterns and regularities. The bodily systems automatically tune themselves to register what goes with what, and what follows what. Attention is itself a skilled and variegated capability. Often when we are searching for something we miss it, or fail to look methodically (what my wife drily refers to as 'having a boy's look'. Though it is not always gender-related: when the cricket ball disappeared into the long grass, as a child I could never find it, and my dad always could.) Wine masters, auctioneers – and batsmen – learn to make sensory discriminations that pass me by. Psychotherapists develop an acute sensibility to tiny telltale clues in their client's posture or voice quality. Rose Rose could sit in a back booth, chatting with her husband and sons, and still monitor with great precision the state of play in the restaurant, knowing almost instinctively whose coffee needed a top-up and who was ready for their bill.[14]

Attention often needs to be patient as well as acute. Karl von Frisch, the man who first described the 'waggle dance' that bees use to tell each other where the pollen is, derived his insights from many hours of patient observation. 'I discovered that miraculous worlds may reveal themselves to a patient observer where the casual passer-by sees nothing at all,' he wrote. Such an attentive attitude towards life can be cultivated (or dulled). Scientists, artists, designers and cooks can all learn to become, like Autolycus in *A Winter's Tale*, a 'snapper-up of unconsidered trifles'. Chemist and author Primo Levi said 'I am very glad that I educated my nose', as a prime tool for identifying chemical compounds. Strolling around town with e.e. cummings, a companion observed 'he would continually be noting down groups of words or scribbly sketches on bits of paper'.[15]

The education of touch plays an important role in many fields of expertise. It is not just readers of Braille who learn to 'see' with their fingertips. Throughout history the goldsmith's assay depended on his experienced palpation of his metal, judging its nature by its feel and consistency. The tailor tells her cloth by its touch, and the cabinetmaker strokes the grain of a piece of cherry-wood both lovingly and cognitively, allowing his imagination to be informed by the whorls and knots he is knowing through his skin.

Deserving of a category of its own is *learning by imitating*. Close observation of a role model serves as a template for one's own tentative actions. Much derided though it has been, the traditional apprenticeship of 'sitting by Nellie' – watching how Nellie does things, and trying to copy what she does – is one of our most powerful learning tools. As we saw in Chapter 9, our brains are full of 'mirror neurons' that innately predispose us to do exactly that. But, as with other forms of attention, we can get sharper or duller at it as we grow up.

And our word-obsessed culture often induces a reciprocal dulling of our powers of observation-imitation. University of California professor Barbara Rogoff has demonstrated marked differences in this regard between children in traditional Mayan communities in Mexico, and American children. In one study, she had an adult ask two children, aged around five, if they would like to learn how to do an origami trick. They would obligingly say yes, but were told that one would have to wait nearby while she explained it to the other. Of interest was the behaviour of the waiting child. The Mayan children sat quietly, fixated on what was going on between the other child and the adult. In a visual presentation by Rogoff, she showed us a 90-second film clip of a Mayan observing child – and it honestly looked like a still photograph. American children, by contrast, very quickly became fractious and fidgety. A week later, the Mayan observers were able to perform the origami trick very well by themselves; the US children had learned almost nothing – because they had not been watching. More literate and numerate the American children may be, but somewhere along the line they had sacrificed other, less intellectual but equally valuable, learning capabilities.[16]

The Siamese twin of observing is *learning by doing*: perception is no use if it does not become a guide for skilled action.

Sensibility and expertise have to be welded together. Remember the poor kitten who saw the world go by, but learned little from the experience if its own legs were not involved in *making* it go by. But learning by doing comes in different forms and covers a range of intentions. In observation-imitation, especially as one's own level of skill develops, the goal is not literally to become a clone of the model. The learning process is far from mindless 'aping'. The body you are observing is different from your own in all kinds of ways – in age, musculature or even gender. The observation gives your body-brain ideas and templates, but it is only through acting them out that you discover where the model fits and where it chafes. Through practice the template becomes customised, dissolving into a skilfulness that inevitably bears your own personal stamp.[17]

Even observation is carried out with an active mind. Marshall Faulk was a legendary running back for the St Louis Rams football team. Like thousands of other virtuosi, as a youngster he spent hours and hours watching his heroes play. As Faulk tells it, he always watched with one burning question in his mind: why? Why did he run there? Why did he not run there? Why is the team hanging back in defence rather than pushing out? As a result he developed not only skill but an unmatched sense of positional play. One of his team-mates commented, 'Marshall has the highest football IQ of any position player I've ever played with.'[18] In real-world learning there is no distinction between the intellectual and the non-intellectual: they weave seamlessly together.

Some learning by doing aims for efficient, reliable automation of a skilled move, and for that there is no substitute for hours of physical practice. Training the neurochemical processes to respond automatically and reliably, on demand, takes time; fine-tuning the body's capabilities operates on a different time-scale from the learning that takes place in a classroom or a business seminar. Just changing a simple habit – such as always choosing fruit for dessert rather than the creamy alternative, until it becomes second nature – takes around two months.[19] The much-quoted figure is 10,000 hours from novice to expert. But this is only half the truth: some people get there much quicker, some never at all, no matter how much time they put in. And one of the key variables is what they are doing with their minds while they are practising. Smart practising involves monitoring your own progress, setting realistic goals, designing your own training regime, knowing when to take breaks and when to push through, picking out the 'hard parts' for special attention, knowing who to ask for what kind of help, and where to place your attention, for example.[20]

But if all you aim for is automation, don't be surprised if you turn out to be an automaton. Fluent expertise needs judgement, flexibility and creativity as well as reliability. So some practice time (and attention) needs to be earmarked for a more playful attitude: not just playing the 'game' (whether that be tennis, surgery or haute cuisine) as well as you can, but playing *with* it. (If you want to see a virtuoso display of precision mixed with playfulness, watch Cristiano Ronaldo at http://www.youtube.com/watch?v=q4y1ypgttwU.) In any skilled activity there is a vital role for 'messing about', trying new tricks, seeing what happens when you force the material (silk, egg white, football) to behave differently. The craftswoman finds time to bend her skills and tools to see what happens when she does. That way she uncovers interesting new problems and opportunities. The two goals of reliability and creativity must be yoked together. As Richard Sennett says, 'Technique develops by a dialectic between the correct way to do something and the willingness to experiment through error. The two sides cannot be separated.'[21]

Tinkering and experimenting spark *learning through imagination*. And again, this comes in different flavours. The controlled version is often called mental rehearsal, and involves seeing yourself in your mind's eye, and feeling yourself in your mind's body, doing something you want to do, better than you currently can. There is a wealth of research now which shows that such mental activity is a powerfully effective adjunct to physical practice itself. Imagination, like emotion, is a halfway house between body and mind – an embarrassment for the Cartesians who want a clean split, but a clear indication of their unity for the rest of us. As we also saw earlier, there are softer or 'low ego-control' versions of imagination in which we let things 'come to us' in a state of relaxed but vigilant reverie. We can get better at slowing our thoughts and watching them unfurl from their embodied beginnings, often catching interesting, shadowy nuances that may have got bleached out as those thoughts evolved into more fully conscious versions.

In practice, these non-intellectual, more embodied ways of learning and knowing blend together with thinking of a more rational and explicit kind. But none of them, certainly not reasoning, is senior to the others. They work together like an experienced jazz combo, sometimes playing solo, sometimes in harmony or counterpoint, sometimes feeding each other lines to be developed in each instrument's characteristic tones.

Actually, many of the learning strategies that I have so far attributed to bodily learning can apply to intellectual learning as well. It just seems, in school at least, that intellectual learning has become exaggeratedly narrow and dull. In English secondary schools especially, there appears to be a pervasive belief that, if learning is to be taken seriously, it must be made dull. This was a problem that struck Seymour Papert. Papert co-founded the world famous Robotics Lab at MIT, and invented Logo, a simple but powerful programming language for children. Early in his career, he worked in a maths classroom in a junior high school in Massachusetts, and every day he had to walk past the art room to get there. The students were carving sculptures out of blocks of soap. They worked on them for weeks. Papert took to stopping by, and was fascinated (and challenged) to observe a depth of engagement, thoughtfulness, creativity and collaboration that he had never seen in maths.

He came to realise that a big part of the difference lay in the fact that the students were working with their hands to craft something 'real' that they could think and talk about as it developed. They were not just 'mindlessly' whittling away at the soap; they were highly present, using the whole gamut of learning methods woven together. They were looking carefully and feeling with their fingers; experimenting and tinkering as they went along; imagining and wondering about new possibilities; and being thoughtful and self-critical too. To call what they were doing merely 'manual work' was to miss completely the richness and complexity – the cognitive sophistication – of the learning they were doing.

Papert saw that Jean Piaget, his old mentor, had got it wrong. As children grew up, they did not move from physical learning through imaginary learning and on to formal or rational learning, leaving the earlier modes behind as they 'outgrew' them. On the contrary, imagining and reasoning *added to* observing and experimenting, making practical learning more and more intricate and powerful. Children were not moving through stages in a linear way; they were developing a richer and deeper *repertoire* of learning instruments that could all play together. That's what those students were learning in the art room, and to Papert the learning experience in maths came to look rather thin beside it. He resolved – very successfully – to find ways to make learning in maths as rich and 'hands on' as it was in the art room down the corridor.[22]

*****

It is only when we look at waiting tables and jointing wood through the already-biased spectacles of rational intellectualism that they appear simple or menial, for those blinkers actively prevent us from seeing the intricacy and intelligence in the work of a waitress or a carpenter. As Mike Rose says,

> We operate with a fairly restricted notion of intelligence, one pretty much identified with the particular verbal and quantitative measures of the schoolhouse and the IQ test. And thus we undervalue, or can miss entirely, the many displays of what the intelligent mind does every day, right under our noses.[23]

Take those blinkers off and we see how interesting and elaborate the minute decision-making processes of these somatic workers actually are. Looked at afresh, many accomplishments of the body, both mundane and esoteric, are shot through with genuine intelligence.

11

REHAB

HOW CAN I GET MY BODY BACK?

Introspective observation is what we have to rely on first and foremost and always.

William James[1]

An embodied approach to human intelligence gives pride of place not to rationality but to sensibility. Reason is a fine tool, like the surgeon's scalpel, good for some specialised tasks but not for everything. However well honed, it is not always the right tool for the job, nor always used skilfully. (Common) sense and sensibility must come first. So, from the embodied point of view, before any training in logical analysis or 'rational thinking' must come a well-developed sensitivity to the processes of one's own body. If, in the process of developing rationality, that sensibility has been lost or muted, intelligence itself is diminished and needs to be rehabilitated. In this chapter we will look at some of the ways in which this can happen. But before we get going, I need to reiterate one important point from Chapter 8.

On the conventional view of mind, we might justify strengthening our body-consciousness on the grounds that some intelligent, interior observer, looking at this visceral data, can be better informed and thus make better decisions. But the embodied point of view on consciousness, as I have developed it, is different. The entire human system is self-organising. There is no 'little person in the head' who does the cognitive heavy lifting: who 'pays attention', 'makes decisions' and 'plans actions'. When aspects of our internal activity become linked to consciousness, this is often because they are currently bound in to the high-level 'dynamic core' where all the different loops of neurochemical activation come together. So the benefit of 'being conscious' of our gut feelings is simply that, when awareness does arise, the underlying pattern of somatic activity is playing its part in the 'central committee'. Its information is being taken into account. The benefit of developing greater 'interoceptive awareness', therefore, is that more of our viscerally embodied feelings and values are 'at the table'. If they are not accompanied by consciousness, it probably means that, though they may well be active, they are not currently engaged in contributing to that top-level democratic decision-making process.

Interoceptive awareness

People vary hugely in how much of their own bodily activity they are routinely aware of, and in how much, if they really try to focus, they can bring into awareness. Stephen Porges has developed a Body Perception Questionnaire that enables you to check your own awareness.[2] It asks, for example:

Are you never / occasionally / sometimes / usually /always aware of:
    Your body swaying when you are standing
    The pressure of the floor on your feet

The speed and depth of your breathing
Swallowing
The strength of your heartbeat
The temperature of your face
The feeling in your stomach
Sensations in your neck and shoulders
The feel of your clothes on your skin
Changes in your voice quality

And there's another questionnaire, the Multidimensional Assessment of Interoceptive Awareness, MAIA, that goes into more detail on the social and emotional aspects of body awareness.[3] The MAIA has items like

'I notice when I am uncomfortable in my body'
'I distract myself from sensations of discomfort'
'I notice how my body changes when I get angry'
'I listen to my body to inform me about what to do'

People vary markedly in how they answer such questions, showing how differently people engage with their own bodies. Some of us are highly sensitive and attentive to changes and activities under the skin; others of us remarkably ignorant. And it makes a difference.

We've already seen how some of these differences in body awareness influence higher mental processes. People who are more sensitive to their heartbeat perform better on the Iowa gambling task, for example. Their greater bodily acuity enables them to register more fully the visceral legacy of previous experience, so this information feeds more accurately into the central decision-making processes that determine the choices that are made. If people's interoceptive awareness is muted, however, they seem to have more trouble with decision-making. People suffering from clinical depression, for instance, are poorer than average at monitoring their own heartbeat – and they also, according to a survey of their therapists and psychiatrists, experience difficulty with decision-making in everyday life. In depression, it is not just that life seems bleak; the withdrawal of attention from both outside and inside deprives sufferers of those vital somatic markers that would normally have prompted and guided their choices and actions.[4]

Similar difficulties are experienced by people with eating disorders such as anorexia, bulimia or 'body dysmorphic disorder' (when the body feels a different size, weight and shape to the way it actually is). Such people are often quite insensitive to the visceral feelings that signal when they are hungry or 'full'. And when our sense of hunger is not underpinned by its normal visceral triggers and anchors we are at greater risk of eating (or not) for a host of reasons unrelated to nutrition. Anorexic individuals may as a result have difficulty sensing their own weight loss, and thus persist in thinking that their body weight is high when it is in fact already low.

There may be several reasons why interoceptive awareness can go up and down, and science has not yet pinned down when and where each of these possible causes applies. It may be that internal systems do partly shut down sometimes, so the neurochemical and muscular activities are themselves less intense and less 'communicative'. This seems to be the case with meditators and others who are able to enter states of deep calm, for example. You might expect such folk to be better than average at feeling their insides, but one recent study found that when they were in this calm, restful state their introspective awareness was no better than normal. The authors of this study suggest that this is because their systems really have shut down so there is just less to feel. There is other evidence that meditators, when they are getting on with life, are indeed more sensitive to their inner workings than a matched sample of non-meditators.[5]

But it is also possible that bodily awareness is reduced when signals in the 'normal' range are not being processed properly as they loop up into the brain, and especially into the areas where all our sources of sensory information come together: the cingulate, the insula and parts of the prefrontal cortex. As we have seen, the brain is perfectly capable of muting feelings that, for whatever reason, it has decided are threatening or inappropriate, so that information is inhibited or expurgated as it unfurls. It can also be blocked from achieving the conditions necessary for it to become conscious, particularly from being linked to the currently active core of self-related activity. In this latter case, the gut feeling may still be playing at full strength, but we don't know it.

We may suffer a drop in our awareness of the bodily self simply because the central machinery is not working properly. If the insula is on the blink, awareness will fade for physiological rather than psychological reasons. Professor Bryan Lask, President of the Eating Disorders Research Society, has shown that this is the likely cause in some cases. His functional magnetic resonance imagery (fMRI) studies show abnormal blood flow to the insula of recovering anorexics, which would make the communication between body and brain slower and less efficient. It could also explain why anorexic women often have difficulty in relating visual, tactile and visceral sources of information about the body to each other. To recover fully, medical treatment, as well as or instead of psychological therapy, may be what's needed.[6]

Then again, it could be that the processing of bodily information is going OK, but there is too much activity in other systems for it to be heard. People who suffer from chronic pain, for example, can have the same trouble making decisions as depressed people, but for different reasons: the physical and emotional clamour of the pain simply drowns out other, perhaps less intense or urgent, visceral voices. In fact, we might all suffer from a version of this when our conscious self is arguing with itself so noisily that we can no longer sense our own bodies. Sometimes this 'loud self' becomes so habitual that the 'still, small voice' of our bodily feelings, values and concerns is shut out of the central decision-making process for long periods of time. (Although such talk used to be New Age whimsy, I think we are now able to put it on a sounder footing.)

Thinking itself could be contributing to this dysfunctional internal racket. In a world that confuses intelligence with rationality, we can be persuaded to think too much. The kind of thinking that is valued and promoted in the Cartesian world is necessarily conscious, and that means it is energetically expensive: it takes up a lot of the bandwidth of the underlying neurochemical systems. So clear, conscious, deliberate thinking can easily overwrite sources of quieter or more fleeting information. We saw earlier that golfers may, under pressure, activate a range of internal voices dispensing 'good advice' and find that their embodied performance – just attending to the ball, the putter and the hole – loses fluency and spontaneity as a result. We can overthink problems and make worse decisions, and be less creative, as a result. People who are thinking hard about problems, especially those that require a creative leap, do worse than those who report periods where their minds 'just go blank'. In one study, students who were invited to choose an art poster for their room were more satisfied with their choice if they had based it on 'gut feeling' rather than on clear, justifiable reasons. People who are weighing up a difficult decision, such as choosing an apartment to rent from a wide selection, make better decisions if they spend a period of time, before making the decision, not thinking about it. In each of these cases, the contribution from the visceral, feeling centres of the body turn out to be crucial, and intelligence deteriorates if that contribution is overridden or drowned out.[7] Numbing down equals dumbing down.

So what can we do – assuming the damage to our complex interior is not irreversible – to smarten up? In the rest of this chapter I will review four methods proven to be effective. The first is biofeedback: training ourselves to be more sensitive to and in control of our insides. The second is meditation, particularly a widely used form called mindfulness practice. The third is a therapeutic process called focusing, which increases awareness specifically of activity in the torso, where all our major organs live. This both increases interoceptive awareness directly and quietens competing forms of noisy activity in the mind. And the fourth is exercise and movement, some, but not all, of which helps to enhance intelligence. Actually there is a fifth form of somatic therapy as well – shifting our *understanding* of human intelligence so our minds make room for our bodies. That is the whole purpose of this book.

Biofeedback

Whether there is any extra damping or not, biological signals are often faint and diffuse. Our language often reflects this vague character: queasy, edgy, out of sorts, off-colour, jittery, in the pink. Our expressions for our overall physical state are regularly indistinct and metaphorical. So it would be useful if we could turn up the volume and the resolution on those faint signals in order, at least, to learn to recognise them more clearly. And we can. There are many machines that can register different aspects of our physiological activity more precisely than we can ourselves, and then play that activity back to us in more vivid ways. You can see the electrical activity of your brain trans-duced into the colours of the rainbow, for example, or the varying conductivity of your skin converted into a rising and falling tone. With that heightened information, you can, often without knowing how you do it, learn to control that signal yourself. That is biofeedback.[8]

Through this methodology you can learn many useful things. You can learn to calm yourself, regain bladder control, control asthma, relieve headaches and reduce chronic pain. You can also learn to increase the coherence of your own biological functioning. Earlier we talked about the healthiness of having a heart that beats (within limits) irregularly, so that it can reso-nate with what is going on elsewhere in the body. If the heart is 'strung too tightly', so to speak, it becomes rigid and unrespon-sive to the other 'instruments' around it. It turns out that, if the heart rate variability (HRV) is displayed as a trace on a screen, people can learn to increase it – they can 'loosen the string' a little, and the body then works better as a whole. It takes on average four to ten sessions of practice to get the hang of this, and then the knack becomes a habit and there is no further need for the machine. Retuning and rebalancing the internal systems improves health; it also enhances skilled performance. Groups of basketball players and wrestlers were given some sessions of HRV training, after which, compared to controls, their speed of movement and reaction was significantly faster, their level of concentration was improved, and the ball players shot more baskets.[9]

But biofeedback can also boost more cognitive forms of intelligence. Children diagnosed with attention deficit hyperac-tivity disorder (ADHD) typically show rather different electro-encephalograph (EEG) patterns from matched controls.

Through neurofeedback they can learn to change these patterns, and this has been shown in a wide range of studies to have a beneficial effect on their ability to control their impulses and concentrate. The ADHD children also show significant gains on a variety of intelligence tests and other tests of cognitive performance. In one study, children given biofeedback training gained 22 IQ points, and this gain was maintained through various follow-up tests over more than four years. Neuroimaging studies showed that the inhibitory functions in the frontal lobes became more like those of non-ADHD children.

Similar benefits have been found with elderly people. Training them to be able to increase a particular aspect of their EEG called the Peak Alpha Frequency led to an improvement in the speed with which they could perform mental computations, and to greater control over their ability to stay on task. Overall, it is clear that biofeedback technology and training can help people to gain greater control over aspects of their bodily functioning that are usually outside of awareness, and that this control brings psychological benefits.

Mindfulness meditation

Some of the same benefits can be achieved without the use of biofeedback machines, simply by training yourself to attend more fully and with greater stability to bodily sensations. Attention is like a muscle that becomes stronger and more controllable over time; apparently it can be developed just like any other habit. One of the most well-known ways of developing this attentional control is through a kind of meditation practice called mindfulness training. In this training, you sit still and quietly, gently attempting to keep your attention focused on a simple bodily sensation such as the rise and fall of your chest and abdomen as you breathe naturally. Despite all the familiar background chatter and distractions of the so-called 'discursive mind', it is possible, with patience and persistence, for that interoceptive awareness to become clearer and more stable, and it can come to function as a kind of anchor that keeps your awareness more strongly focused in the present and reduces the length of time you spend ruminating over past and future.[10]

Studies of mindfulness practitioners have shown how this shift in attentional habits is mirrored in the functioning of the brain. Norman Farb and his colleagues at the University of Toronto have identified two complementary networks in the brain's frontal lobes. One involves pathways in the medial prefrontal cortex and the language centres, and this network keeps tabs on our overall portfolio of personal projects, interests and disappointments. The other involves the lateral prefrontal cortex, the insula and the anterior cingulate, and focuses on what is going on in the present moment both within and outside the body. Seasoned meditators show less activity in the medial system and more in the lateral. They have upped their 'presence of mind' and reduced the time spent anxiously planning, reviewing and rewriting the scripts for achieving their goals and righting the wrongs they (think they) have suffered.[11]

In general, it seems as if mindfulness training increases the quality of information that arrives at the central core of neural decision-making from both internal and external sources. One recent study has shown that even something as 'physiological' as the functioning of the immune system is improved by mindfulness training.[12] The immune systems of meditators respond more effectively to infection than do those of non-meditating control subjects. Meditation also improves the quality of communication between as well as within the body's sub-systems. It increases the efficiency and comprehensiveness with which these sources are interwoven. The different loops become functionally better integrated – and this leads to better decision-making and so, in an important sense, to greater intelligence. The better I am informed, and the more fully that information is integrated, the better my system computes 'the best thing to do next'. Behavioural evidence supports what we know about what is going on in the brain. Meditators are better at sustained concentration: they are less susceptible to distraction from whatever their 'task at hand' may be.[13] And their brains seem to react faster and more fully to quickly changing events. Most of us, for example, are subject to the so-called 'attentional blink', in which seeing one thing reduces the likelihood that we will notice another event that follows close on its heels. Meditators show less attentional blink than the rest of us.[14]

One striking example of the benefits of mindfulness comes from a study of a group of long-term meditators taking part in what has come to be called the Ultimate Game. In the two-person version, A can choose how much of a $20 gift they will offer to share with B. The catch is: if B rejects the offer, they both get nothing. Rationally, it makes sense, if you are the second person, to accept even low offers: at least you get something rather than nothing. That's what the meditators are inclined to do. The meditators' 'gratitude', you might say, tends to outweigh their indignation that someone else is getting more than they are. Most people, however, 'cut off their nose to spite their face' by rejecting offers of less than around 20 per cent. We are unable to resist the urge to punish someone we think is behaving unfairly, even if we lose out as a result.[15]

Interestingly, this same study also monitored brain activity, and found a relative increase, in meditators, in activity in the posterior part of the insula. According to the researchers, this may signify a greater ability to feel our visceral responses, but without letting those feelings automatically grab the steering wheel of our actions. This gives quite a sophisticated boost to our intelligence: we are able to take our visceral responses fully into account, but to weigh them in the context of wider considerations of both our own and other people's best interests.

So this increased ability to be fully present to *all* the information, including our own core values (which for most of us includes altruism or 'being a good person'), means that it is not just intelligence in an abstract sense that benefits; our wider social and personal well-being increases too. A patient who had spent some time in a form of body-oriented psychotherapy, learning to pay closer attention to her bodily sensations and reactions, sums up this subtle step-change in full, real-world intelligence very clearly. The training, she said,

> has done a lot to help me be calm when I'm in a stressful situation. So that when the stress arises – when you get to the airport and your plane is delayed nine hours and there's no flights and no hotels and everyone else is sort of screaming – I don't join in that any more. Now I can just see, 'Oh, I'm feeling a bit agitated . . . time to start breathing!' And realising that you have the ability to respond rather than react and the degree to which all of us are on automatic pilot most of the time. It's like, 'No, you have options here; you can choose how to respond to this situation.'[16]

The distinction she makes between responding and reacting is an important one, and it brings us back to System One and System Two (Chapter 5). Reacting, as she uses the term, relies a lot on habits and assumptions. For example, the current situation is quickly assimilated into one of the dominant emotional modes – anxiety, indignation or sadness. Responding is the ability of the body-brain to restrain this tendency to leap to conclusions and react in stereotypical ways, and to gain a fuller, more nuanced appreciation of the situation, in all its aspects, before actions are selected and initiated.

Focusing

Bodily awareness is made up of all kinds of sensations – and people may be sensitive to one kind and not to another. Meditators, as we saw, have been found to be more sensitive than average to touch on their skin, but no better than average at judging their heart rate when they are in a very quiet and relaxed state. Athletes and dancers can become very attuned to the feeling of their muscular bodies, but not to their emotional bodies (a dissonance very effectively dramatised by Natalie Portman in the film *Black Swan*). Developing interoception is a many-sided job, and different kinds of training may be needed to develop different areas.

Sensitivity to what is going on in the major cavities of the body – the throat, the chest and the abdomen – is one of the most important areas when it comes to developing 'intelligence in the flesh'. And here it is worth mentioning a specific practice called 'focusing' that has been developed to train exactly this aspect of our bodily awareness. Back in the 1960s, Eugene Gendlin and his colleagues at the University of Chicago were trying to identify the 'magic ingredient' in counselling and psychotherapy that made some clients feel they were making productive progress, while others were spinning their wheels. After analysing hundreds of tape recordings of therapeutic sessions, they found it. It wasn't the school of the psychothera-pist, or even their personality or 'bedside manner'. What made the difference, in the majority of cases, was whether clients spontaneously talked about their troubles in a particular way.[17]

The stuck clients (to put it very crudely) tended to trot out tales of woe fluently and rapidly. They had their well-rehearsed story – and they were sticking to it. But, strikingly, the ones who were getting the most benefit talked much more hesitantly. It was as if they were trying to find just the right words to express the complex truth of their predicament, and were listening carefully to their own formulations to see if they did justice to this truth. Moreover, it turned out that the touchstone of this truth was not a cognitive response but a visceral one. Clients were not rationally appraising what they had said, but allowing their spoken words to resonate with what they were feeling in their bodies, especially in the torso, where the major organs and the visceral core are situated.

Gendlin called this pre-verbal, embodied sensation the 'felt sense'. The successful clients were in touch with this felt sense and were checking to make sure that the essential meanings were preserved as it unfurled into words. When they were, Gendlin discovered, clients experienced a bodily feeling of relaxation – 'Ah yes, that's it. That's exactly it.' The felt sense would then change, and deeper understanding, integration or reconciliation would emerge. In a picturesque image, Gendlin once described this process as like 'listening to the child in your chest'. When the child feels accurately heard and appreciated, it can, so to speak, stop squirming and grizzling, and unwind. It feels as if an uncomfortable blockage, like indigestion, has been dissolved, releasing a fuller, more integrated feeling of inner flow and harmony.

Gendlin's genius was to discover that this knack can be learned by almost everyone. (Having done several focusing workshops I can testify to this.) You ask yourself a very general question – 'What's this whole thing with my workplace / my mother / my fear of being alone about?' – and then hold your attention in your torso and wait to see what sensations form there. It might be a sinking feeling in the pit of your stomach, a feeling of being unable to breathe, a pricking in the back of your throat, or perhaps a slight tensing of the shoulder muscles (or a hundred other things). You learn the art of slowing down the unfurling process – the differentiating of this deep concern into more precise words, feelings or gestures – so that it doesn't cut corners, or quickly replace the actual with the 'normal', but preserves the authenticity and the complexity of the felt sense as it gradually wells up. And that process brings insight and a feeling of forward movement. Gendlin says: 'Underneath your thoughts and memories and familiar feelings, you can discover a physically sensed "murky zone" which you can enter and open. This is the source from which new steps emerge. Once found, it is a palpable presence underneath.'[18]

Try it. Be patient: it works. But it works best in communication with someone else. The interpersonal bodily resonance that we looked at in Chapter 9 is a very useful medium for developing this greater sense of integration and unfolding. In a telephone seminar in 2011, Gendlin explained why.

> Focusing is a way to access your bodily knowing. Your body picks up more of the other person than you consciously can. Your body also puts out more of yourself than you intend or than you know is visible. Others often react to that rather than to your conscious message. With a little training you can get a feel for your bodily knowing of what is going on.[19]

As the therapist tunes into and resonates with the client's total embodied state, so the client, with some conscious direction, can learn to do the same. A growing body of research is tracking the neurochemical correlates of these intra- and inter-personal resonances. Not surprisingly, the prefrontal cortex, insula and cingulate are all implicated again.[20]

You could say that psychotherapy is concerned with mental health rather than cognitive performance – though the distinction clearly breaks down in this new view of real-world intelligence. For example, Gendlin has more recently discovered that the same kind of patient inward attention contributes significantly to the process of creative thinking. Gendlin has developed a kind of coaching protocol in which one person can help another to take the inkling of a novel idea, which they do not initially know how to express very well, and slow down its unfurling so that a genuinely satisfying expression of this idea can gradually evolve and come together. He has dubbed this 'Thinking at the edge' (or TATE, for short). Like therapeutic focusing, the process starts with a general question: 'In your professional field, or in your life, what do you "know" and cannot yet say, that wants to be said?' Your partner then helps to guide your sensing and thinking with a kind of subtle, patient midwifery, until you begin to find new forms of words (or other symbols), the *mots justes*, that 'do justice' to the original inkling. Though this research is still in its infancy, there are several encouraging studies of the use of TATE in various managerial, academic and educational contexts.[21] (Indeed, writing this book has been a long process of trying to find good ways of putting new inklings into words.)

Exercise and movement

Running! If there is any activity happier, more exhilarating, more nourishing to the imagination, I cannot think of what it might be . . . The mysterious efflorescence of language seems to pulse in the brain in rhythm with our feet and the swinging of our arms.

Joyce Carol Oates

Mens sana in corpore sano – a sound mind in a sound body – may be a saying of dodgy provenance, but is there any evidence that physical exercise really makes us smarter? We all know that exercise is good for our health and well-being. But what about our intelligence? The evidence is that exercise does seem to benefit thinking – though not always. Simply becoming more muscular or developing athletic expertise does not have any direct effect on our intelligence. In fact, heightened levels of mental and physical control can enable one to dampen and ignore feelings of distress or exhaustion – as, again, with Nina, the obsessional ballerina in *Black Swan* (and some top athletes and performers).

Yet physical movement can contribute positively to intelligence in at least three complementary ways. The first view sees exercise as increasing the physiological performance of the body-brain by, for example, improving the supply of oxygenated blood to the brain, or simply by getting the organs to work more efficiently.[22] But there are more direct effects on intelligence as well. Exercise certainly affects brain areas such as the hippocampus that are strongly implicated in learning and memory. Regular running makes more neurons grow in these areas, and has a beneficial effect on the speed and effectiveness with which people learn. In one study, people learned new words 20 per cent faster after a burst of intense physical activity. Exercise is especially useful for protecting our minds from the effects of ageing. The hormones involved in regenerating neurons decline as we get older, and exercise offsets that decline.

It's also the case that the connectivity of the brain decreases as we age; the different circuits become slowly more disconnected from each other. This effect too is offset by exercise.[23]

But exercise affects thinking in young people too. Aerobic exercise especially makes a difference in tasks that require inhibitory control by the frontal lobes: planning, sequencing and prioritising, and keeping on track towards a current goal despite temptations and distraction. More specifically, practising complex physical skills such as dribbling a football, *in the context of exercise that makes you tired*, seems to accelerate the development of those frontal lobe faculties even further. Even regular bouts of less skilful physical activity during lessons have been shown to boost achievement in reading, spelling and mathematics – and thus, not surprisingly, to show increases in IQ.[24]

Secondly, the role of exercise can also be seen as vital for keeping internal communication between organs and sub-systems at an optimal level. The body is a mass of interconnecting, interlocking, intercommunicating sub-systems. As in any relationship, these communications vary not just in content but also in quality. If my wife and I stop communicating, we no longer have a marriage. We no longer resonate with and shape each other in an elastic and evolving way. Being well tuned, well tempered – in good temper with ourselves and the world – turns out to be a key ingredient of intelligence. If a part of me becomes too flaccid, it cannot pick up the detail of what it is sensing. And the same is true if it becomes sclerotic. When we become hard-hearted, we lose our sympathetic resonance with the world – and especially with the people – around us. A chronic loss of empathy is tantamount to a loss of intelligence.[25]

If we become sedentary and/or obese, or succumb to certain states of 'mental illness', it may well be that we literally lose a kind of constitutional physical elasticity that is important for the systemic resonance of the body's components. As yet there is little research in this area, and more would be of great interest. It seems likely, for instance, that the accumulation of 'hard fat' around the vital organs, as in the 'apple shaped' form of obesity, would reduce inter-organ communication as well as bringing the well-known health risks. The fat could cause compression of sensory nerves, restriction of fluid-borne neurochemical messenger molecules, or literal damping of physical resonance. One might also wonder about the effects of pharmaceutical drugs such as statins, blood-pressure regulators and weight-loss pills on the effectiveness of the body's internal communications.

The third benefit comes from exercise that increases our interoceptive awareness (and not all exercise has that intent or that effect). T'ai chi is a prime example of exercise where bodily awareness is explicitly cultivated. It is a venerable, originally Chinese, exercise regime that consists of slow, flowing bodily movements requiring detailed coordination of head, eyes, arms, torso, pelvis and legs. The movements are very precise but, to begin with, quite unfamiliar and so require a high degree of concentration and subtle body awareness to master. In a review of well-conducted evaluations of the cumulative effects of t'ai chi, Peter Wayne (US Scientist of the Year 2013) and his colleagues at the Harvard Medical School found significant increases in bodily awareness, with clear benefits for both mental and physical functioning. Elderly practitioners of t'ai chi are less likely to fall (and suffer the consequences of bruising and bone fracture that can follow), for example, because they feel the sensations of their feet on the ground more fully and accurately. Ironically, because they are more present, and less preoccupied (as many elderly people are) with a fear of falling, they fall less often. And the functioning of their internal systems improves more generally, so that there are real benefits to memory and problem-solving as well. These benefits are significantly larger than any that accrue from forms of physical exercise that do not focus on the cultivation of bodily awareness.[26]

Hatha yoga also stresses bodily awareness, and research suggests that it can have short-term as well as long-term benefits for the way we think and solve problems. Neha Gothe and her colleagues at the University of Illinois compared the effects of a mere 20 minutes of yoga practice with the same amount of time spent in conventional stretching exercises (which do not emphasise awareness). Their subjects were female college students who had no previous experience of yoga, t'ai chi or any similar form of exercise. Five minutes after a period of exercise or yoga, they were given tests of their ability to focus on a task and ignore distractions. Performance on these tests was significantly better when they had just had the yoga session than after conventional exercise. So the good news is that you do not have to practise the disciplines for months before you begin to experience the cognitive benefits; a mere 20 minutes will give you a 'quick win'.[27]

Dance training, too, requires the dancer to develop a high degree of bodily awareness – though the emphasis may be more on the muscles, and the positioning and shaping of the body in space, than on the more visceral aspects of interoception. So we might expect a mixed picture when we go looking for cognitive benefits of dance. Some of the yoga studies, for example, used a dance class as a control group, and found little cognitive benefit of dance as compared with Hatha yoga. Peter Lovatt, however, has found a direct effect of dance on the way people think. Peter is both a professional dancer and an academic psychologist at the University of Hertfordshire. With his colleague Carine Lewis, Peter gave novice dancers 15 minutes of either free, improvisational dance, or of highly structured dance steps, and then tested them on their ability to solve two different kinds of puzzle. One required creative insightful thinking, the other more logical and analytical thinking. The effect of the dancing on their problem-solving was quite specific. The improvisational dancers showed improved performance on the insight problems but not on the analytical problems, while the structured dancers showed exactly the reverse effect. Although a preliminary study, it does suggest, intriguingly, that mind and body can mirror each other's mode very quickly – as we would expect if they actually are, and are felt to be, two sides of a coin.[28]

*****

Overall, it is pretty clear that we can indeed improve internal communication between our different organs, with a corresponding increase in mental agility and performance. Our ability to learn, remember, pay attention, solve problems and engage with others are all dependent on the quality of the communication within and amongst that complicated circuitry. And that quality can be recovered and enhanced through different kinds of exercise and therapy. So what, in the light of this knowledge, might I be inclined to do differently?

At a very superficial level, I might try some 'power posing' before an interview, using body posture to induce a corollary state of confidence. I might exercise more, in the belief that 'being in better shape' would benefit not just my health and longevity, but my mind as well. I'd think of the goal not as becoming fitter or firmer, but as toning my system so that it is better tempered, more in tune with itself and its surroundings. I wouldn't be aiming to run faster or to defy the sagging and wrinkling of age, but to thrum more sweetly and respond more intelligently to the constant plucks of the innards and the 'outtards' that compose myself. And I might practise slow movements, or even sitting still, in order to learn to listen and feel these plucks and throbs more fully – not so that a disembodied mind can be 'better informed' (for I accept now that this ethereal governor doesn't exist), but so that every member of the corporeal choir can contribute its particular voice more fully to the central chorus out of which my intelligence emerges. I'll practise noticing my heartbeat, the rise and fall of my breathing, my blinking and the faint background clamour of tensions and itches, and tremor in my muscles, so that no part of me, if I can help it, is excluded from the moot.

But what about the deeper effects of embodiment on our very sense of self? And how would society need to change if its members experienced themselves as fully embodied? These are the questions we will address in the final chapter.

12

THE EMBODIED LIFE
SELF, SPIRIT AND SOCIETY

Gone is the central executive in the brain – the boss who organ-ises and integrates the activities of multiple special-purpose sub-systems. And gone is the neat boundary between the thinker (the bodiless intellectual engine) and the thinker's world. In place of this comforting image we confront a vision of mind . . . not limited by the tenuous envelope of skin and skull.

Andy Clark[1]

So here I am. I am a body. I am a body that is not so much a lump of too, too solid flesh, topped with a mind, but rather a swirling flux of information streams. These streams have both internal (e.g. gut) and external (e.g. eyes) origins, though they interweave so quickly and thoroughly that it is impossible to separate them out and say 'who I am *really*', or what 'the real world' looks like, or 'who started it'. 'Who I am' is already deeply influenced by messages coming in through eyes and ears, nose and tongue and skin. I am, through the body, intricately embroiled in the environment, both physical and social, and it in me. And the world, as it appears to me, is already deeply tinted and flavoured by my own values and history. We are built to be chameleons, taking on the colours of the world in which we find ourselves; and also decorators, designing and annotating the world to which we then respond.

Even to speak of 'having a body', or of something occurring in or to 'my body', is to misrepresent the situation; it is to insert the thinnest but most insidious of wedges between the body and a hypothetical owner or driver. Even to write – as I was just tempted to – of 'the world around me' or 'my environment' is to presume that Me and the World I Inhabit are separate, or at least separable, and this is simply incorrect. That's what the science of embodiment tells us. However deeply embedded these dualistic habits are in our language and thought, they are crude and often misleading approximations to our real nature. Our common sense about ourselves is not innocent or transparent. Identifying with the image we see in the Cartesian mirror has consequences, and not all of them are good, or even neutral.

I am a body that unfurls sometimes into mind-like thoughts and experiences of various kinds, as well as into bodily actions like reaching for the salt or changing the subject. Some of these experiences I treat as being (more or less reliable) depictions of the World Around Me. Some of them I interpret as being reactivated records of my past, or images of desired or dreaded futures. Some of them are diffuse, like the feeling of being off-colour, or of having forgotten to do something. Some materialise as well-formed trains of thought, or blinding insights; others seem to lurk in the wings of consciousness: mental marginalia such as inklings and hunches. And I treat some of these more shadowy backdrops to consciousness as aspects of my self: the feeling of witnessing a performance, or intending an action, or resisting a thought. Sometimes these fronds of consciousness co-occur, and then I have the impression of a self thinking, or planning an action, or rehearsing a move, as if one frond were the cause or controller of another. But really, the 'I' is as much an upwelling from the interior as the thought or the action itself. I am a body–mind–context constellation, ever changing and ever welling up. And so, I think, are you.

What happens if we try to take this alternative view of ourselves ever more seriously; if we move along a path from understanding it to accepting or agreeing with it, to believing it, to embodying it? Could 'being an embodied and extended system' become not just an idea or a spur to daily walks but a lived experience – and, if so, what difference would it make to the way we perceive and act and think? And beyond that, what difference would it make to the social world and its institutions and practices? Would embodied education, embodied law, embodied politics, embodied religion, even embodied medicine, be different from the schools, law courts, senates, churches, mosques and hospitals that we know? To do justice to these questions would take another book, but in this final chapter I am going to let my hair down a little and play with some possibilities.

Embodied lifestyle

Seeing that the dynamic communication system of the body is vastly more complex than the conscious mind can capture might lead me to adopt a humbler tone in the way I think about my health. Health is a function of the entire embodied and embedded System, but my conscious mind is often preoccupied with a superficial jumble of folklore, old wives' tales and the media's latest tips and 'findings' about aspirin or statins. (Consciousness often incorporates a clutter of knowledge and opinion that bears only a loose relationship to the felt truths of bodily experience, yet may stride forward and take precedence.) So I might decide it is smarter to take time to tune in to the subtle groundswell of bodily feelings and sensations, rather than espouse the latest nostrum (whether it emanates from an Ayurvedic guru or a professor of oncology).

And, to complement this, I might develop greater acuity about exactly when and where conscious, deliberate reasoning is the best tool for the job, and when it isn't. I'll notice that to think and argue explicitly requires the world I am thinking and talking about to be pretty simple and clear-cut. Rational thinking is like juggling: you can only keep so many balls in the air (and lawyers and professors can cope with not many more than your average bear). If the world is amenable to such simplification, well and good – but mostly it is not, and then the application of deliberate reasoning becomes Procrustean: much of what is complex and intertwined has to be lopped off or ignored to fit the format of rational comprehension and discourse. So it is no wonder that such debates – like the one I watched on television last night about the pros and cons of Scottish independence – regularly become a slanging match between apparently intelligent people who deploy every trick in the rhetorical book to make a 'telling point'. Truth and its pursuit are the first casualties of such rhetorical warfare. As Wittgenstein might have said, whereof one cannot speak clearly and logically, thereof one must bumble – speaking hesitantly, humbly and perhaps poetically. Or not at all.

Intelligence often grows out of a patient willingness to say 'I don't know' and abide in uncertainty. So, in the face of a decision that is important but not urgent, I might try to learn to slow down, pay attention to the faint forming of a felt sense in my midriff or my throat, and see if I can allow the unfurling or the welling up to happen at its own pace, so that I don't skip vital steps and foreclose on the meaning prematurely. In general I might take a growing interest in intuition: signals that emerge from my dark depths in the form of physical promptings, inklings, hunches and images. I do not have to 'buy' them unquestioningly, any more than I have to buy the economic forecasts produced by high-powered computerised reasoning; but I might discover that they have a greater value and validity than the Cartesian model allows.

I might grow to be less troubled by some of the apparently weird or inconsequential stuff that wells up into consciousness, trusting that (provided my Sub-Systems are reasonably well integrated) I am not 'going crazy'; that these puffs of black smoke are just by-products of my system cleaning and conditioning itself, in exactly the same way that my printer, at seemingly random moments, emits a range of grunts and clanks as it engages in a mysterious operation which it calls 'aligning'. There is a calm in accepting Pascal's dictum that 'the heart has its reasons of which reason knows nothing' (and so do the larynx, the lungs and the gut).

Knowing that I automatically become part of a Super-System with the people around me, I might become more gentle and tentative about passing judgement and attributing blame. Both your and my behaviour emanate from the dynamics of this larger system and cannot be accurately attributed to either one of us alone. So, as you and I embody this appreciation, our disputes take on a different tone and we are both participants and explorers, trying to see why this intricate Super-System which we created, and which now creates us, behaved in the harsh or apparently self-defeating way it just did.[2] If I see myself as a ghostly mind, it is easy to feel that I am an isolated and self-contained individual: a localised bubble of consciousness. But if I am a body, first and foremost, then I am already deeply connected. Once I really get that mind is the conscious accomplice of the body, not its governor, I can't avoid seeing myself as an aspect of the wider body social and body politic.[3] As cells are to the physical body, so people are to the larger 'corporations' to which they belong. Embodied implies embedded.

I may generally become more open to – less surprised by – my variability as I move from context to context. The Cartesian model installs in me the sense of an unwavering core that I take for The Real Me, within which, and up to which, I have to live. Whatever is inscribed within the charmed circle of this 'identity', that is not just who I *am* but who I *have to be*, and so various forms of sclerosis must ensue. If (like the sclerotic heart) I insist on marching to my own tune, rather than rocking and rolling with the shifting currents of the Super-System of which I am a part, that's not integrity, it's stupidity. There is no need to be at odds with myself whenever I catch myself acting 'out of character'. If my actions and experiences arise, moment by moment, from the whole body–mind context that underpins me, why should I be alarmed if, on some occasions, I manifest differently from 'usual'? Finding myself in a novel situation, I am bound to surprise myself. It is perfectly natural.

If the essence of my personhood is not just logical thinking and abstract articulation, my very sense of self will shift. Reasoning and dispassionate debate are relative cultural newcomers and they grew out of deeper, older forms of intelligence that were intimately connected with bodily capacities, bodily needs and bodily feelings, and with the challenges and affordances of the material world. Perhaps I'll grow to feel more of a fluid unity, and identify less with the cacophony of competing voices in my head.

And finally, dying might be easier if I deeply felt myself to be an embodied System within wider Systems: a transient manifestation of a confluence of forces, many unknown and/or beyond my control, rather than that solitary bubble. Giving up the comforting illusion that there is a Happy Hunting Ground in the sky where I shall play peacefully for eternity, I might feel unafraid of being, as the mystics say, simply a dewdrop slipping back into the ocean, in need of no more comfort than a deep acceptance of coming home.

So much for the individual experience of being an embodied, embedded being. But what about public life?

Education

Education in general would have to change if a unified body–mind–context view of ourselves were to supplant the opponent dualisms of Cartesian Man. (Many feminist writers have pointed out that the disembodied intellect is very lopsidedly masculine.) As I said earlier, schools rest on a Cartesian view of intelligence. The ability to manipulate abstract and Platonic entities like 'prepositions', 'parabolic curves' and 'Newton's Laws of Motion' counts for more, in school, than the ability to make an old engine run sweetly or to decorate a house beautifully. So Maths and English and Physics are called 'hard subjects', are given more time, and carry more weight on your child's CV than Home Economics and Metalwork. Despite repeated attempts to redress the balance, 'vocational' or 'technical' education is still widely seen as what you do if you are not 'bright enough' to do well at English, Maths or Science. Attempts to raise the esteem of Hairdressing or Motor Mechanics by bulking them up, like supermarket chickens, with watered-down injections of theory, are ineffective, and deeply misguided, because they reinforce the very assumption that needs to be challenged: that mind-stuff – 'book-learning' – is necessarily harder and better than

body-stuff. The education of touch and smell, of visceral awareness and subtle grip, doesn't form a big part of the school curriculum. If a teacher asked her pupils if they were aware of their
heartbeat, and whether they would benefit from being more
aware, she would be thought very weird.

It was claimed by Sir Christopher Frayling, when Rector of
the Royal College of Art in London, that the original '3 Rs' of
schooling in Regency England were Reading, Reckoning and
Wroughting (making things), but that, around 1807, Sir William
Curtis ditched wroughting and elevated writing to an 'R' of its
own, thus further downgrading the manual and practical.[4] In
commenting on this retrograde move, Professor Bruce Archer,
also of the Royal College of Art, noted that 'modern English has
no word, equivalent to literacy and numeracy, meaning the
ability to understand, appreciate and value those ideas which
are expressed through the medium of making and doing. We
have no word, equivalent to Science or the Humanities, meaning
the collective experience of the material culture.' Yet many
scientists value highly this physical and material sensibility. I
once heard the venerable scientist and bioengineer Heinz Wolff
talking about his working methods. He was discussing the
important role that old-fashioned metal Meccano had played in
his development as a scientist, and suggested with some passion
– and a strong German accent – that 'it is as important for
young people to become manipulate as it is to be articulate'. So
maybe we can look forward to an enlightened society in which
manipulacy is talked about and valued as highly as literacy and
numeracy.

This hierarchy of esteem cannot be justified on the grounds
of utility. Being able to recite the rules and categories of
grammar does not make you a better writer. Learning how
to solve simultaneous equations does not make you a better

all-round thinker and problem-solver. And, beyond basic arith-metic, statistics and probability, more complex maths can best be learned when you need it (as I and many others have done). There is no practical justification for inflicting trigonometry on everyone. The fact that a few youngsters love it, and that it perfectly fits the Platonic mould, are inadequate arguments for making it universal. The hierarchy persists because the mind–body split, and its inequities of esteem, persist. Not just inequities but iniquities. Because Descartes had no way of understanding the cleverness of bodies and brains, and we still follow his model, thousands of young people who are good with their hands and feet, but not so good with equations or apostrophes, have been led to think that their talents and interests are second-rate, demanding, perhaps, of hard work and practice, but not requiring much in the way of genuine intelligence.[5]

Politics and law

Cartesian education induces changes in young people that carry through into their working lives and thus can influence the way our public institutions work. Let me give you just one (slightly controversial) illustration from the United Kingdom. In his books *The Making of Them* and *Wounded Leaders,* psychotherapist Nick Duffell reports detailed case studies of the psychological effects of boarding school on children's visceral and empathic sensibility – especially if they are as young as seven or eight when they are sent there. It is a British tradition, which continues to this day, to send boys (and some girls) to boarding schools, many of whom will emerge as cultural leaders in the media, the judiciary, the armed forces, the City and, especially, in government. The current British

Prime Minister (David Cameron) was seven when he became a boarder, and around two-thirds of his cabinet have a similar background.

Their overt education is highly intellectual and analytical. Years are spent honing the skills of comprehending and critiquing knowledge, and of constructing convincing arguments on paper and in debate. In seminars and tutorials at university these skills will be refined even further. This intensive apprenticeship in the etiquette of rationality produces what the French call a *déformation professionnelle*: not just a set of skills but an attitude towards life that depends on and privileges those skills.[6] Duffell claims that this bias is accompanied, in many boarders, by a serious dampening of interoception. 'Children survive boarding by cutting off their feelings and constructing a defensively organised self that severely limits their later lives,' he says. The pressures of being a young boarder (as I was myself) may require you to learn how to pretend convincingly and continuously that you are not upset, scared or homesick, for to do so risks looking ungrateful to your parents, who are spending lavishly to buy you this privileged education, and 'babyish' to your peers, who will eagerly pounce on anyone who displays such weakness. I still remember with shame doing nothing to stop a night-time posse of ten-year-olds peeing on the bed of that term's designated runt, a poor boy we called 'Barmy' Wright – with Barmy in the bed. Many of us were too frightened it would be our turn next to say anything. As many studies of deception have shown, the best way to lie to others is to lie to yourself; self-deception – denying the very experience of those tell-tale feelings of weakness – is the trick to learn.

The emphasis on sport in such schools can cut both ways, as far as the development of intelligence goes. Attitudes of resilience, fortitude, hard work and practice, as well as camaraderie and teamworking, are self-evidently valuable resources. Yet – as with the ballet dances Nina – that determination may depend on neglecting 'inconvenient' messages from the body, whether of muscle strain or tiredness, or of other competing values or needs. Visceral processes (as we saw in Chapter 6) can be dampened, through the use of muscular tension, and disconnected neurochemically from the central somatic processes that bind together feeling, action and perception. And the habitual attenuation of bodily processes can have a powerfully negative effect on social and emotional intelligence, as well as on cognition. It would seem a gross exaggeration to claim that we have been accepting, as leaders, people systematically trained to lack the deepest kind of intelligence there is. Yet a recent letter to *The Observer*, signed by many leading psychiatrists, psychotherapists and neuroscientists, expressed exactly this concern.[7]

*****

The Cartesian model shapes personalities, and also, over time, shapes professional practices themselves. Legal and political proceedings claim to respect only rational argument based on objective evidence – though they routinely admit the use of a wide variety of rhetorical tricks that bear no relationship to rationality. At High Table and in the Debating Chambers of government and law, decisions depend a great deal on who can mount the most persuasive (or least refutable) argument, or twist the arms of colleagues most successfully, and not on who can bring us closest to the truth or to wise judgement. Speaking from the heart, perhaps with a large reservoir of relevant experience, in a way that touches people and reminds them of deeper values and considerations, has had no formal place in such proceedings, and is often treated as embarrassing.

The inadequacy of cleverness is regularly on display in the ritualised rhetorical warfare of the law court. Everyone has to pretend that non-verbal signals, gut feelings and years of experience are of no account, while sophistical wordsmiths try to bamboozle jurors into making their preferred decisions. Points are scored when a seasoned expert is tripped up by a tricky question. Decisions that make no sense from an embodied perspective – such as whether someone was in their 'right mind', or 'knew what they were doing' – are treated with a farcical degree of seriousness. Just as swearing on a bible should carry no weight in a post-supernatural world, neither should much of what is now an anachronistic judicial process.

In other societies, less infatuated with the ideal of dispassionate reason, important decisions are approached differently. In traditional Maori culture, the weight that is given to someone's thoughts on a matter depends largely on their *mana*, and this reflects their experience, their reputation and their personal and physical bearing. Someone with high mana speaks with an integrity that seems to emanate from a deeper, more embodied centre of knowing than merely the articulate intellect. In Maoridom there is faith that a gathering can be helped towards a wise, humane and inclusive resolution of an issue by the words of such a person more effectively than they can by clever verbal combat. The frequent fallibility of the adversarial judicial system, and the equally frequent silliness and superficiality of political debate, have certainly given me and many others cause to doubt the Cartesian systems we have inherited.

If wise judgement turns out to be very different from the lawyerly arts of sophistry and spin, who we consider fit to govern us will have to change. And these changes would obviously be resisted by those who, as Machiavelli put it, have done well under the old dispensation. Those who (like me) are good at reading, writing and reasoning will be rattled by the removal of the taken-for-granted linkage between their intellectual talents, their social esteem and their income.

People would need to be governed and motivated differently if prowess at reasoning were no longer considered the epitome of intelligence. One of the unquestioned values in many contemporary societies is social mobility. The desire to 'better yourself' – to 'do better' than your parents – underpins both education and the economy. But the magnetic pull of 'betterment' is usually towards mind-work and away from body-work. Around the world, the kinds of work that people aspire to, and those that are best paid, tend to be those that seem to require mental rather than physical skills. Shuffling paper in a clean shirt in a run-down government office is presented as a more honourable, because more *intelligent*, way of earning a living than working on the family farm or caring for young or old relatives. As Matthew Crawford, author of *The Case for Working with Your Hands*, says, 'it is a peculiar sort of idealism that insistently steers people towards the most ghostly kinds of work'.[8] There are whole societies that pride themselves on outsourcing any form of manual labour, from making coffee to building skyscrapers, to immigrant workers, and in which the insignia of 'success' are fine white garments that demonstrate the ultra-cleanliness of the micro-world you inhabit. In such societies, the idea of embodied intelligence could be completely, and dangerously, revolutionary.

Medicine

Even medicine has come to see the body as a collection of parts or symptoms which can be treated individually as one mends or replaces the parts of a machine. As I age, I can take pills for my blood pressure, pills to lower my cholesterol, pills for a headache, pills to shrink my prostate gland, pills to ward off an attack of gout and pills to try to deal with that persistent cough. My sneaking suspicion that these conditions might be interrelated, that a more systemic approach might be called for, is not something that my GP can address. The medical and pharmaceutical industries' ability to understand how all my different drugs might interact is almost non-existent. They talk of 'side-effects' as if these were somehow technical problems they just haven't quite managed to fix yet, rather than systemic effects: desperate attempts by a whole body to deal with a complex pharmaceutical onslaught it does not physiologically recognise and for which it has evolved no armoury. Conventional medicine has hardly begun to explore what it grandly calls 'poly-pharmacy'. Even the gruesome term 'multi-morbidity' – suffering from several conditions at once – betrays the medical profession's inability to think systemically about the body. Many elderly people may be suffering from just one thing, old age, a common condition with multiple symptoms, which contemporary 'Western' doctors, for all their sophistication, simply do not know how to conceptualise other than as a collection of separate disorders with a pill for each. Old age is not an acceptable cause of death on an English death certificate; the doctor has to latch on to a single specific thing – in my father's case, 'Parkinsonism' – out of dozens of interlinked changes, to make death conform to this fragmented medical model.[9]

Even the body of the physician is treated as an irrelevance. She has her intellectual knowledge and her technological resources; what more could she want? The Cartesian approach to medicine has no way of incorporating good old-fashioned 'clinical judgement' or of admitting the validity of other, more bodily, ways of knowing that do not rely on numbers and machines. And in a litigious world, if something goes wrong you are much better off with a computer print-out of a lab test than a defence of 'thirty years of experience'. The research is clear: experience manifesting as intuition has real validity, and it can also be wrong – just as rational analysis of a spreadsheet can give a stupid answer. The fallibility of 'gut feelings' is a reason not to ignore them but to tread cautiously and to try to understand better whose intuition is more likely to be valid when, and why. But you can't teach clinical intuition in a one-semester course, or assess it with a multiple choice test.

Even in this small discussion, by the way, we can see how the Cartesian worlds of medicine, education and the law lock together into a self-reinforcing system that validates reason and neglects the body. A specific example concerns midwives who, for the reasons cited, are increasingly reluctant to act on their intuition. An experienced midwife may well have a hunch about when a home birth is going well enough, despite a complication, to continue, and will judge if and when the ambulance needs to be called. But increasingly she may override her intuition that all will be well and call the ambulance 'just in case'. The evidence is that midwife-assisted home births are at least as safe as, and often safer than, hospital birth, and that, despite all the fancy machinery, there has been no overall improvement in birth outcomes over the last thirty years. The widespread use of foetal heart monitors during labour, for example, has resulted only in more Caesarean operations, not in better outcomes. An authoritative review in *The Lancet* concluded:

> For low-risk mothers, there is a good case for a return to the traditional method of intermittent auscultation [listening to the internal sounds of the mother's body with a stethoscope]

with its lower false-positive rate, lower incidence of intervention, and opportunity for greater contact between the maternity care staff and the mother.[10]

In an embodied world, intuition born of experience would have its proper place, and medical students would be taught more about both the benefits and the fallibilities of their own intuition: when to heed it, and how, over time, to improve its quality. They would also be trained more effectively in the old-fashioned arts of inspecting, listening to and, very importantly, touching ('palpating') and smelling a patient's body and its products. A great deal can be learned by listening carefully to the sound quality of a baby's cry or of an old man's creaky knees, by smelling a patient's breath, or by inspecting the consistency and colour of someone's urine or stools.[11] The idea that such intimacy with your patent is distasteful and, worse, unreliable, is a modern aberration that medicine would do well to overcome. It is not that machines don't tell you useful things; of course they do. But they offer only one source or type of information, and there are other, more somatic ones that it seems just plain stupid to ignore.

Screens and the body

Our emotional and social intelligence suffers if we ignore our bodies. Digital communication is fine, and fun, but if we cannot touch and smell the people we are with, a good deal of value is lost. A study found that American college students are less empathic than they were just a few years ago. They are less likely to agree with statements like 'I often have tender, concerned feelings for people less fortunate than me', and 'I sometimes try to understand my friends better by imagining how things look from their perspective.' The authors of this study note that there could be several reasons for this. Much digital media content, from internet porn to casual 'sexting', is aggressive and disparaging. It is harder to connect with and be influenced by other people's distress if you have become cumulatively desensitised, and also if you cannot see the immediate hurt in their posture or hear the tremor in their voice. One of the authors of the study, Edward O'Brien, puts it like this. 'The casual relationship people have with their online "friends" makes it easy to just tune out when users don't feel like dealing with others' problems and emotions. As these social media relationships consume more and more of our time, it's easy for this online behaviour to bleed into everyday life.'[12] It could also be that our hand-held devices encourage a disposition towards fast reaction, and thus limit our bodily involvement. Antonio Damasio has found that our higher emotions such as empathy or compassion require biological processes that are inherently slow. It takes seconds, not milliseconds, for such feelings to germinate and unfurl within us. So chronic speediness could be reducing our emotional register in damaging ways.[13]

Just adding emoticons doesn't cut it; we are not evolved to have the same visceral reaction to ☺ or ;-(as we do to photos of real people smiling and weeping. Clever designers are doing their best to reconnect our smartphones with our bodies. You can buy a Bluetooth-enabled Hug Shirt that will respond to certain messages on your phone by contracting and giving you a friendly squeeze. But – forgive me if I sound old-fashioned – I don't think it would be quite the same somehow . . .

A lot of companies now have had to ban smartphones from meetings as a result of the destructive effects of what they call 'continuous partial attention'. Smart devices are a mixed blessing, it seems; they can augment our intelligence, and they can also make us dumber. The more time we spend looking at our screens, the less we may be watching other real-life human beings reacting in subtle ways in real time to the world around them, and gradually learning to tune our own resonances to those frequencies. As Mark Bauerlein, Professor of English at Emory University, says, 'The digital natives improve their adroitness at the keyboard, but when it comes to their capacity to "read" the behaviour of others, they are all thumbs.'[14]

The current obsession with digital technology raises a number of such concerns, and they are coming at us so thick and fast that research cannot document which are real problems and which are new forms of perennial moans about the decline of the young. Screen life is sedentary. Yes, you can do physical exercise or play virtual golf with your Wii, but the vast majority of screen life requires fingertips and eye muscles only. The body's massive musculature is trivially engaged. Does that matter? Screen life is sensorily impoverished. It offers sight and sound, but – with the exception of the tapping of keys (and the Hug Shirt) – nothing solid and smelly, tickly and tasty in the way that real life is. Computers lead us to concentrate more and more on the two evolutionarily younger senses, sight and hearing, and to withdraw attention from the others. Are we being numbed as well as dumbed down by our machines? And, when so much information can be accessed and stored at the click of a mouse, does the opportunity to offload memory have consequences for the way we think and the quality of the ideas we generate? Does screen life risk encouraging us to be merely well informed rather than creatively intelligent? Time will tell; but there is enough evidence to suggest that these are serious questions, not to be lightly dismissed as a nostalgic knee-jerk reaction of the elderly.[15]

A new materialism

One symptom of dissatisfaction with screen life is a resurgent interest in physical pursuits and accomplishments. There is, I sense, a New Materialism around: one which is not about shopping and displays of conspicuous consumption, but about the pleasures of making and moving, from quilting to skateboarding, and much in between. Indeed, it seems that the more the digital world takes hold, the stronger, for many of us, is the reaction to get back from the virtual to the substantial, from the symbolic to the concrete: from mind to body.

Stanford University, I've heard, has instituted a compulsory course for new students in architecture, engineering and medicine that requires them to strip down and rebuild a bicycle, or build a model aeroplane from scratch. Of course, all Stanford students arrive with impressive school grades, but these have often been found to mask a deep lack of physical problem-solving ability and common sense. Intellectual accomplishment is no guarantee of practical intelligence. In a similar vein, one of the most heavily oversubscribed courses at MIT is MAS.863. It is called 'How to Make (Almost) Anything'. This is its description.

> This course provides a hands-on introduction to the resources for designing and fabricating [things]: machining, 3-D printing, injection moulding, laser cutting; PCB layout and fabrication; sensors and actuators ... wired and wireless communications. This course also puts emphasis on learning how to use the tools as well as understand how they work.

Amongst the 'brightest' students in America there is, apparently, a huge hunger for working with physical material and making real things that really work.

More broadly, the Maker Movement is rampant in many countries of the world, providing free access to both traditional and digital workshops (called Hackerspaces or FabLabs) for people who want to make anything from a bird-box to a dancing robot. FabLabs are available in cities across the planet from Lafayette to Milan to Shanghai. Makers are also putting pressure on manufacturers to make things that are easier to repair. Mister Jalopy, the pseudonym of a maker guru in Los Angeles, has developed a 'Maker's Bill of Rights' (the full manifesto can be found at www.makezine.com). It includes:

- Meaningful and specific parts lists shall be included with everything
- If it snaps shut, it shall snap open
- Special tools are allowed only for darn good reasons
- Individual components, not entire sub-assemblies, shall be replaceable
- Ease of repair shall be a design ideal, not an afterthought

Makers are fond of quoting Marge Piercy's evocative poem 'To be of use'. It includes the lines:

The work of the world is common as mud.
Botched, it smears the hands, crumbles to dust.
But the thing worth doing well done
has a shape that satisfies . . .
The pitcher cries for water to carry
And a person for work that is real.

Gever Tulley's 'Tinkering School' is a summer camp in the US. He guarantees to anxious parents that they have no need to worry: their child *will* come home covered in bruises and scratches. His much-watched TED talk is called 'Five Dangerous Things You Should Let Your Child Do'. They include playing with fire, owning and using a penknife, and making and throwing spears. If we don't let children do these things, Tulley argues, they grow up alienated from their physicality – and they don't develop an intelligent appreciation of danger when it comes their way. He shows slides of Inuit toddlers using razor-sharp knives to cut whale blubber – having already learned key maxims of safety such as 'Always cut away from you' and 'Don't press too hard'. Tulley now runs Tinkering workshops for software designers at Adobe HQ in San Jose to help them get closer to the hands-on experience of their customers. He was brought in after the Adobe VP for design, Michael Gough, saw the effect of Tinkering School on his own children. Bill Burnett, executive director of the product design course at Stanford, says, 'A lot of people get lost in the world of computer simulation. But you can't simulate everything . . . All your intelligence isn't in your brain. You learn through your hands.'[16]

In all the excitement about the affordances of the digital world, we must not forget that it has limitations, and it has its casualties. Yes, you can save money by outsourcing the interpretation of X-ray scans to India, but there is much skilled, satisfying, necessary work that cannot be done down a wire or via a satellite – because it requires skilled muscles and refined sensibility. You might be able to use your smartphone to record a TV programme while you are waiting in the departure lounge for your homeward flight, but a Beijing plumber is no use to you if your London sink is blocked. A stylist in Melbourne cannot cut your hair. A childminder in Phnom Penh can't look after your baby. While many jobs can be automated or outsourced, the need for physical strength and intelligent hands is not going to die out.[17]

The casualties? Automation is, without question, depriving many people of the visceral satisfaction of baking a good loaf of bread or assembling a real car. Sociologist Richard Sennett has studied the changing fortunes of a bakery in Boston. Back in the 1970s the work was hard, you had to get up early, people got burnt, but there was a sense of communal pride in a delicious, beautiful batch of focaccia. There was craft: the flour varied in texture, so an experienced baker, feeling the flour in his hands, would make an adjustment to the amount of oil added or to the baking time. By the late 1990s, there were a lot of part-time workers who hardly knew each other, and computer-controlled machines that just had to be fed bags of 'croissant mix' when the appropriate icon blinked. The environment is safer now, but it is craftsman-hostile, because human craft is neither needed nor possible, and so no pride ensues. All the intelligence has been appropriated by a machine. But if the machine breaks down, there is hardly anyone left who actually knows how to make bread. One old Italian man said to Sennett: 'I go home, I really bake bread: I'm a baker. Here? I just push buttons.' As Sennett's work powerfully attests, this loss of pride in craft knowledge, and of a satisfying sense of belonging to a skilled community, is a personal tragedy, but it is also, writ large, a social, even a global issue. The demeaning of work, and the estrangement from first-hand contact with real material, has demonstrable costs. Greater 'productivity' comes at a high price – though it is not one you can easily factor into your spreadsheet.[18]

If I can come back to the point at which I started this book: this cost is largely unacknowledged because the Cartesian world does not recognise the sophisticated intelligence of manual work. Matthew Crawford, in *The Case for Working with Your Hands*, nails this point. 'The trades suffer from low prestige, and I believe this is based on a simple mistake. Because the work is dirty, many people assume it is also stupid.'[19] The more body, the less intelligent is an idea that still pervades public consciousness. But the dimensions *intelligent–unintelligent* and *intellectual–practical* do not map neatly on to each other.

Physical work can demand the highest intelligence – and not only (as we saw in Chapter 10) in the esteemed professions like medicine, engineering or architecture. Peter King is both a shop-class teacher and a practising car mechanic in Colorado and a friend of American educator Art Costa. Costa has done brilliant work on unearthing the 'habits of mind', as he calls them, which underpin intelligence in many different spheres. King loves his work as a mechanic but he also feels rather sheepish about it. He recognises that all the habits of mind that Costa talks about – problem-solving, checking accuracy, persisting in the face of difficulty, creative thinking and so on – are present in his work. Yet he still catches himself categorising 'mechanics' in general as unintelligent people. He wrote to Costa:

How hypocritical is it of me to think that mechanics are stupid when I'm one myself? My passion is fixing my car, making it go faster and better. So how could I think badly about mechanics? Well, it's that little thing called 'peer pressure'. My parents, friends and the majority of people look down on people who fix cars. So I look down on myself. I hide my hobby like it was a crime.

People don't realise the massive amount of problem-solving power it takes to fix someone else's mess. Don't get me wrong; there are definitely bad mechanics. That's why I fix my car myself. But all those habits of mind you talk about – those 'characteristics of intelligent behavior' – they are all there in what I do.

We are all 'mechanics' in a way. It's just that some of us get our hands greasy.[20]

Conversely, the world is full of mediocre mind work. Matt Crawford describes the time he was employed writing brief summaries of academic articles, mostly on subjects he knew precious little about: anything from classical philology to microbiology. He thought he would be using his brain, but it turned out he was to use a formula that meant he did not require any understanding of what he was reading. His quota, after a year, was 28 articles a day. That's 14 minutes per article. He was a 'knowledge working' skivvy, on $23,000 a year. British journalist Giles Coren recently described his own similar work. 'I just sit and Google. It's terrible. I wish I was a fireman.'[21]

Real-world intelligence

If you want to teach a computer to play chess, or if you want to design a search engine, the old model is OK. But if you're interested in understanding real intelligence, you have to deal with the body.

Rolf Pfeifer[22]

So finally, let me come back to my guiding word, intelligence. It's a tricky word because it is part hypothetical entity, part value judgement, and part a history of acrimonious debate. Of the last I shall say nothing here: that history is already well digested and rehearsed.[23] Its residue is a widespread view of intelligence that is too narrow and too intellectual; too focused on, and too in awe of, the kinds of problem-solving that professors of philosophy do, and, reciprocally, neglectful and disdainful of the intricate, situated thinking that happens every day in motorcycle repair shops, hospital wards and restaurants. Abstract reasoning is a useful element of intelligence but it is not the whole thing, and the ability to deploy such reasoning judiciously, in the mêlée of a busy café or shop, requires powers that the professor may well lack. IQ tests commonly fail to predict the intelligence of people's performance in real life, and where abstract reasoning could be used, it is often found to be more elegant and reliable to make use of 'tricks of the trade' rather than working things out from first principles.[24] We should remember, too, that the conception of what counts as intelligent differs widely between societies. In some societies, intelligence is inextricably linked with kindness and social grace, for example. In others, it includes the ability to remember the complex myths of the society and to retell them with accuracy and flair.[25]

As a starting point for the development of a science of embodied intelligence, let me offer this. Intelligence is getting things done that matter to you. It is finding good resolutions of those three sets of factors: your concerns, your capabilities and your circumstances. And to do that well, you need your body, and you need the kind of broad, detailed integration of its messages that gives rise to conscious awareness. In routine situations we rely, quite rightly, on habit and precedent. It is intelligent of me to operate on automatic pilot most mornings as I prepare to go to work. Intelligence is what allows effective, economical, elegant and appropriate interaction with the world. It is intelligent to accumulate a wide repertoire of such routines.

And it is intelligent to detect nuances of situations that might signal whether they are routine or not. One is often going to 'guess wrong', go with habit, and find that the habitual response is inadequate and something more creative or custom-ised is required. Often the momentary conjunction of Needs, Deeds and See'ds does not fit any familiar mould, and then effective action requires a form of intelligence called 'thinking on your feet' or 'floundering intelligently'. And this is different from IQ. We saw earlier that Google is not impressed by people's track records of success, but is equally sceptical of high IQs. Laszlo Bock, the senior vice-president in charge of 'people operations' – the head of HR – says: 'For every job the No. 1 thing we look for is general cognitive ability, and it's not I.Q. It's *learning agility*. It's the ability to process on the fly.' Behind the ability to learn quickly lies what Bock calls 'intellectual humility'. You have to be able to give up the knowledge and expertise you thought would see you through, and look with fresh eyes. People with a high IQ often have a hard time doing that. They are certainly no better than average at tolerating uncertainty, or being able to adopt fresh perspectives.[26]

Learning agility relies, especially in situations of emergency, disruption or seeming chaos, on finding a way through what had seemed to be intractable. It involves integrating and reconciling all the different pulls and pushes, and this needs careful orchestration by those areas of the frontal lobes that specialise in prioritising, sequencing, keeping on task and dealing with conflicts and glitches. In tricky situations many processes have to interlock and self-organise, modulating activity throughout the rest of the brain and the body, to achieve the best resolution possible. Secondary concerns and activities get muted or deferred; a particular subset of affordances is highlighted; actions that might be useful are fired up and ready to go; attention is swivelled to potentially relevant parts of the world; parts of the memory networks judged to be relevant – possible precedents and resources, for example – are primed.

This network of self-triggering inhibition and regulation is often referred to in psychology as the Central Executive, or Working Memory, but, as we explored earlier, it does not involve a separate kind of intelligence that sits outside the rest of the body-brain; it just operates (principally) at the junctions where all the loops of information come together. These processes do not inhabit a separate 'place' in the brain where things are 'sent' for special attention. Indeed, intelligence does not reside in any single cognitive faculty – certainly not in the specialised tool called 'conscious, disciplined reasoning'. It resides in systemic processes that involve resolution, reverberation, integration and balance. Intelligence is the orchestrating and conducting of a whole ensemble of influences which include, essentially, those of the body; and, through the body, those of the external world. Lumping all these processes into two baskets called System One and System Two doesn't get us very far.[27]

*****

A lot has been written in the last twenty years about different *kinds* of intelligence. We have had emotional intelligence, practical intelligence and 'bodily-kinaesthetic' intelligence, along with a host of others.[28] Many overlap a little with the territory of this book. But there is one key difference: these 'intelligences' are usually presented as part of a larger portfolio of separable psychological faculties. The most well known of these systems, Howard Gardner's theory of Multiple Intelligences (MI), offers eight different kinds of intelligence, of which the bodily and the emotional are two. The intelligences are presented as distinct, complementary and equally valuable, and people can be described in terms of their profile of aptitudes and preferences across the range of intelligences on offer. This refinement is a major improvement on the monolithic idea of intelligence that underpinned IQ.

The view from the new science of embodiment, however, suggests that these intelligences are not separate, and they are not of equal value. It is not that some people have a lot of 'musical intelligence', while others are high on 'logical-mathematical', and that makes for a rich and diverse world. My claim is more radical than that. It is that practical, embodied intelligence is the deepest, oldest, most fundamental and most important of the lot; and the others are facets or outgrowths of this basic somatic capability. Emotional intelligence is an *aspect* of bodily intelligence. Mathematical intelligence is a *development* of bodily intelligence. To identify 'bodily-kinaesthetic intelligence' just with what top gymnasts and artisans have is to miss the fact that, at a deeper level, this one intelligence is in fact the root system on which all the others depend. I don't want to flatten out the hierarchy of esteem that puts Pure Mathematics at the top and Woodwork at the bottom; I want to turn it on its head. A revised understanding of the relationship between body and mind is absolutely key to the development of intelligence per se.

Last word

I think it is time to reclaim the concept of intelligence from the abstract world of disembodied symbols and propositions, logical arguments and rigorous deductions, and proclaim its wider relevance to the challenges and complexities of everyday life. To deal well with life's demands requires a full body – not just for getting around and implementing actions, but because a well-integrated, well-tuned, highly resonant body is itself the organ of intelligence. The brain plays an important part in that integration, allowing loops of information from the skin and the spleen, the hands and the heart, the gut and the gullet to be brought together in fruitful discourse. But without all those loops carrying fast-changing information about what is possible and what is desirable, and without the constant conversation between all the far-flung outposts of the body, the brain would not be intelligent at all. It is only as good as the intelligence it receives. The condition of my body, and of my awareness of its humming, shimmering activity, constantly modulates my ability to be smart.

And through the body, we are deeply connected with and constituted by the world around us. The tools and resources we use literally become incorporated into the body's working definition of itself. Our bodies actually vibrate with each other so that individual Mes begin to dissolve into a larger resonant system called Us. Embodied cognition teaches us to notice that we are much more ecological and social than the Cartesian doctrine has led us to believe.

Somewhat similar images of the human body, its connectedness and its capacity for intelligence have been with us for millennia, of course. You can find such sentiments in the writings of poets, sages and physicians down the ages. But only now do they have the imprimatur of science, and in a global culture that reserves a special place of respect for scientific knowledge, that is important. I do not want this book to give any succour to lazy New Age thinking that muddles up grains of truth with bucketloads of magic and mythology. I have been at pains to point out that gut feelings and intuitions have some important validity, but they are no more immune from interrogation than are the pronouncements of a prophet. But if this book has helped to deepen and enrich our understanding of what and who we are, and of a kind of intelligence that has more to do with practical wisdom than with intellectual cleverness, I shall be well pleased.

NOTES

Chapter 1

1. Richard Rorty, *Philosophy and the Mirror of Nature*, Wiley-Blackwell: Oxford, 1981, p. 239.
2. Please don't think that I am making any claim to the moral high ground here. Writing a book is about as cerebral and sedentary as it gets. And I do indeed have 'bike ride?' pencilled in my diary for the Sunday after next. I'll return in the last chapter to a discussion of why understanding has so little purchase on behaviour.
3. Ballet-dancers, concert pianists and other practitioners of 'high culture' are exempted from this generalisation – though even in school, dance, drama and music sit only just above PE in the hierarchy of esteem. And, in the case of ballerinas at least, they are not supposed to look as if they sweat and strain.
4. Ambiguity intended, by the way.
5. Andy Clark, *Being There: Putting Brain, Body and World Together Again*, 1997, Bradford/MIT Press: Cambridge, MA, p. 180.

Chapter 2

1. John Bannister Tabb, 'Brother Ass and St Francis of Assisi', in *Truth*, Tate Publishing: London, 2014.
2. For a much richer treatment, see Roy Porter, *Flesh in the Age of Reason*, Allen Lane: London, 2004.
3. Which is what, perhaps, gave dramatist Tom Stoppard the idea for his play *Jumpers*, which features a department of academic philosophers who are also amateur gymnasts.
4. David Young, '*Mens sana in corpore sano?* Body and mind in Ancient Greece', *International Journal of History of Sport*, 2005, 22(1), 22–41 is the source of this and the following quotations.

5. E. R. Dodds, *The Greeks and the Irrational*, University of California Press: Berkeley, 1951.
6. Paul's first Letter to the Corinthians: I Cor. 9: 27.
7. William James, *Some Problems of Philosophy*, Harvard University Press: Cambridge, MA, 1979, p. 34.
8. I'll leave, for the moment, the fact that, if they are unrepairable they are broken *up* . . .
9. Quoted in F. Summers, 'Dualism in Descartes: the logical ground', in M. Hooker (ed.), *Descartes*, Johns Hopkins University Press: Baltimore, MD, 1978. I am aware that some recent philosophical thought questions whether Descartes was in fact the 'Cartesian dualist' he has been painted. If I have done Descartes an injustice I apologise, but nothing here hangs on whether the real man was or wasn't a dualist. For the arguments, see Gordon Baker and Katherine Morris, *Descartes' Dualism*, Routledge: London, 1995. Thanks to Jonathan Rowson for this.
10. See http://www.bje.org.au/learning/judaism/ethics/bioethics/body.html.
11. René Descartes, *Discourse on Method*, ed. F. E. Sutcliffe, Penguin: London, 1968.
12. George Lakoff and Mark Johnson, *Philosophy in the Flesh: The Embodied Mind and its Challenge to Western Thought*, Basic Books: New York, 1999.
13. As Sir Ken Robinson says in his much-watched TED talk, 'Do schools kill creativity?', http://www.youtube.com/watch?v=iG9CE55wbtY.
14. Roy D'Andrade, *An Introduction to Cognitive Anthropology*, Cambridge University Press: Cambridge, 1995.
15. Jerry Fodar, *The Language of Thought*, Harvard University Press: Cambridge, MA, 1979.
16. Peter Medawar, 'Is the scientific paper a fraud?', in *Induction and Intuition in Scientific Thought*, Methuen: London, 1969. Paul Feyerabend, *Against Method*, Verso: London, 2010.

Chapter 3

1. Antonio Damasio, *Self Comes to Mind*, Heinemann: London, 2010, p. 36.
2. Jeffrey Deaver, *The Roadside Crosses*, Simon and Schuster: New York, 2009.
3. Daniel Wolpert, 'A moving story', *CAM Magazine*, 2013, 66, 35–37. For a more elaborate version of the same argument, see Rodolfo Llinás, *I of the Vortex: From Neurons to Self*, MIT Press: Cambridge, MA, 2002, Chapter 1.
4. Rodney Brooks, Foreword to Rolf Pfeifer and Josh Bongard, *How the Body Shapes the Way We Think: A New View of Intelligence*, MIT Press: Cambridge, MA, 2007, p. xv.
5. Steve Collins, Andy Ruina, Russ Tedrake and Martijn Wisse, 'Efficient bipedal robots based on passive-dynamic walkers', *Science*, 2005, 307, 1082–1085.
6. Pfeifer and Bongard, op. cit.
7. BigDog is a product of Boston Dynamics, originally a spin-off from MIT and now owned by Google. For a paper on BigDog, go to http://www.bostondynamics.com/img/BigDog_IFAC_Apr-8-2008.pdf.

8. Pfeifer and Bongard, op. cit., p. 20.
9. Raymond Tallis, *The Hand*, Edinburgh University Press: Edinburgh, 2003. I have drawn heavily on Tallis in this section.
10. A. Hernandez Arieta, R. Katok, H. Yokoi and W. Yu, 'Development of a multi-DOF electromyography prosthetic system using the adaptive joint mechanism', *Applied Bionics and Biomechanics*, 2006, 3, 101–112.
11. Llinás, op. cit.
12. I'm paraphrasing Damasio (2010), op. cit., p. 41.
13. Ary Goldberger, 'Is the normal heartbeat chaotic or homeostatic?', *News in Physiological Science*, 1991, 6, 88–91. I am grateful to Dr Stephan Harding for providing me with this reference, and to his (and my) late colleague at Schumacher College, Professor Brian Goodwin. The background to this systemic view of the body is well discussed in Goodwin's book *How the Leopard Changed its Spots*, Weidenfeld and Nicolson: London, 1994.
14. J. Andrew Armour, 'The little brain on the heart', *Cleveland Clinical Journal of Medicine*, 2007, 74(1), 48–51.
15. Francisco Varela and Antonio Coutinho, 'Immune networks: getting on to the real thing', *Research in Immunology*, 1990, 140, 837–846.
16. E.g. Candace Pert, *Molecules of Emotion*, Scribner: New York, 1997.
17. Michael Gerson, *The Second Brain*, Harper: New York, 1998.
18. For a great overview of systems thinking, see Fritjof Capra and Pier Luisi, *The Systems View of Life: A Unifying Vision*, Cambridge University Press: Cambridge, 2014.
19. I'm borrowing here from two of Andy Clark's wonderful books. Andy Clark (1997), op. cit. and Andy Clark, *Natural Born Cyborgs*, Oxford University Press: Oxford, 2003.

Chapter 4

1. Ambrose Bierce, *The Enlarged Devil's Dictionary*, Penguin: Harmondsworth, 2001.
2. Oliver Sacks, *The Man Who Mistook His Wife for a Hat*, Summit: New York, 1985.
3. An 'afferent neuron' is a nerve that sends information towards the brain. 'Efferent' nerves head outwards and downwards from the brain.
4. M. Björnsdottir, I. Morrison and H. Olausson, 'Feeling good: on the role of C fiber touch in interoception', *Experimental Brain Research*, 2010, 207, 149–155.
5. Kevin O'Regan and Alva Noe, 'A sensorimotor account of vision and visual consciousness', *Behavioral and Brain Sciences*, 2001, 24(5), 939–1031.
6. Ira Hyman, Matthew Boss, Breanne Wise, Kira McKenzie and Jenna Caggiano, ' "Did you see the unicycling clown?" Inattentional blindness while walking and talking on a cell phone', *Applied Cognitive Psychology*, 24 (5): 597–607; Arien Mack and Irvin Rock, *Inattentional Blindness*, MIT Press: Cambridge, MA, 2000. If you haven't seen it yet, try the demonstrations at www.invisiblegorilla.com.

7. Hugo Critchley, 'Electrodermal responses: what happens in the brain', *The Neuroscientist*, 2002, 8(2), 132–142.
8. Anthony Bechara, Hanna Damasio, Ralph Adolphs and Antonio Damasio, 'Deciding advantageously before knowing the advantageous strategy', *Science*, 1997, 275, 1293–1295.
9. R. Held and A. Hein, 'Movement-produced stimulation in the development of visually guided behavior', *Journal of Comparative and Physiological Psychology*, 1963, 56(5), 872–876.
10. A large and very technical literature has emerged over that last twenty years on this issue, and I am not going to be able to do it justice. Key resources include the following: Susan Hurley, *Consciousness in Action*, Harvard University Press: Cambridge, MA, 2002; Dave Ward, 'An action-space theory of conscious vision', PhD thesis, University of Edinburgh, 2008.
11. Bernhard Hommel, 'Action control according to the Theory of Event Coding', *Psychological Research*, 2009, 73, 512–526.
12. Wolfgang Richter and nine others, 'Motor area activity during mental rotation studied by time-resolved single-trial fMRI', *Journal of Cognitive Neuroscience*, 2000, 12(2), 310–320; Rob Ellis and Mike Tucker, 'Micro-affordance: the potentiation of components of action by seen objects', *British Journal of Psychology*, 2000, 91(4), 451–471.
13. Richard Abrams, Chris Davoli, Feng Du, William Knapp and Daniel Paull, 'Altered vision near the hands', *Cognition*, 2008, 107(3), 1035–1047.
14. Dennis Proffitt, 'Embodied perception and the economy of action', *Perspectives on Psychological Science*, 2006, 1(2), 110–121. It seems as if this effect only holds when the motivational factor is of evolutionary origin. Some classic studies which purported to show that poor children saw coins as bigger have been discredited.
15. Emily Balcetis and David Dunning, 'Wishful seeing: desirable objects are seen as closer', *Psychological Science*, 2010, 21, 147–152.
16. Marcel Kinsbourne and Scott Jordan, 'Embodied anticipation: a neuro-developmental interpretation', *Discourse Processes*, 2009, 46, 103–126, p. 103.
17. For a good summary of the arguments about and evidence for prediction coding, see Andy Clark, 'Whatever next? Predictive brains, situated agents and the future of cognitive science', *Behavioral and Brain Sciences*, 2013, 36, 181–253.
18. E.g. Karl Friston, 'The free-energy principle: a unified brain theory?', *Nature Reviews: Neuroscience*, 2010, 11, 127–138.
19. Guy Claxton, 'Why can't we tickle ourselves? On the psychophysiology of sensation attenuation', *Perceptual and Motor Skills*, 1975, 41, 335–338. Some readers may have been wondering: if we cannot tickle ourselves, how come masturbation is not a completely thankless endeavour? Though empirically untested, I suspect that the brain's sexual system must find it more difficult to perform the cancelling than in the case of tickling. Further research may or may not be needed.

Chapter 5

1. Iain MacGilchrist, *The Master and his Emissary*, Yale University Press: London, 2010.
2. The idea of loops is well developed in Damasio (2010), op. cit.
3. M. Lockey, G. Poots and B. Williams, 'Theoretical aspects of the attenuation of pressure pulses within cerebrospinal fluid pathways', *Medical and Biological Engineering*, 1975, 14, 861–869; Robert Provine, *Curious Behavior*, Belknap Press: Cambridge, MA, 2012.
4. William Tyler, 'The mechanobiology of brain function', *Nature Reviews: Neuroscience*, 2012, 13, 867–878.
5. Ana Fernandez and Ignacio Torres-Alemán, 'The many faces of insulin-like peptide signalling in the brain', *Nature Reviews: Neuroscience*, 2012, 13, 225–239.
6. John Cryan and Timothy Dinan, 'Mind-altering micro-organisms: the impact of the gut microbiota on brain and behaviour', *Nature Reviews: Neuroscience*, 2012, 13, 701–712.
7. Stephen Collins, Michael Surette and Premysl Bercik, 'The interplay between the intestinal microbiota and the brain', *Nature Reviews: Microbiology*, 2012, 10(11), 735–742.
8. Hugo Critchley and Neil Harrison, 'Visceral influences on brain and behavior', *Neuron*, 2013, 7, 624–368.
9. Llinás, op. cit. Damasio (2010), op. cit., calls maps 'images', but I find this invites us to import all kinds of dubious and unhelpful connotations of consciousness, so I'll stick to 'maps' here.
10. Antonio Damasio, Paul Eslinger, Hanna Damasio, Gary van Hoesen and Steve Cornell, 'Multimodal amnesic syndrome following bilateral temporal and basal forebrain damage', *Archives of Neurology*, 1985, 42(3), 252–259.
11. A very accessible tour of these maps, and how they are made, is provided by Sandra Blakeslee and Matthew Blakeslee, *The Body Has a Mind of its Own*, Random House: New York, 2008.
12. Hugo Critchley, 'Psychophysiology of neural, cognitive and affective integration: fMRI and autonomic indicants', *International Journal of Psychophysiology*, 2009, 73, 88–94.
13. I'm drawing heavily in the next few paragraphs on the work of Bud Craig. He has been challenged on some of the detail, but I am certain that the general tenor of his proposals must be right. See: A. D. Craig, 'How do you feel? Interoception: the sense of the physiological condition of the body', *Nature Reviews: Neuroscience*, 2002, 3, 655–666; A. D. Craig, 'How do you feel – now? The anterior insula and human awareness', *Nature Reviews: Neuroscience*, 2009a, 10, 59–70; A. D. Craig, 'Emotional moments across time: a possible neural basis for time perception in the anterior insula', *Philosophical Transactions of the Royal Society, Series B*, 2009b, 364, 1933–1942.
14. Meaning: 'I feel as full of energy and optimism as a crate of newborn chicks or ducklings'.
15. Blakeslee and Blakeslee, op. cit., pp. 115–116.

16. Craig (2009b), op. cit., p. 1934.
17. Anil Seth, Keisuke Suzuki and Hugo Critchley, 'An interoceptive predictive coding model of conscious presence', *Frontiers in Psychology*, 2012, 2, 1–16.
18. Rodney Brooks, 'Intelligence without representation', *Artificial Intelligence*, 1991, 47, p. 142.
19. Marcel Kinsbourne, 'Awareness of one's own body: an attentional theory of its nature', in Jose Luis Bermudez (ed.), *The Body and the Self*, Cambridge, MA: MIT Press, 1995; Alva Noë, *Out of Our Heads: Why You Are Not Your Brain, And Other Lessons from the Biology of Consciousness*, Hill and Wang: New York, 2009; Chris Frith, *Making Up the Mind*, Blackwell: Oxford, 2007.
20. Herman Hesse, *The Glass Bead Game*, Holt, Rinehart and Winston: London, 1943; 1968 edn.
21. Salvatore Aglioti, Joseph DeSouza and Melvyn Goodale, 'Size-contrast illusions deceive the eye but not the hand', *Current Biology*, 1995, 5, 679–685.
22. L. F. Barrett and Moshe Bar, 'See it with feeling: affective predictions during object perception', *Philosophical Transactions of the Royal Society, Series B*, 2009, 364, 1325–1334.
23. Much has been written about this so-called 'executive function' of the brain. Good syntheses can be found in Marie Banich, 'Executive function: the search for an integrated account', *Current Directions in Psychological Science*, 2009, 18 (2), 89–94; and Elkhonon Goldberg, *The Executive Brain*, Oxford University Press: New York, 2002.

Chapter 6

1. Mark Johnson, *The Meaning of the Body*, University of Chicago Press: Chicago, 2007, p. 66.
2. Daniel Goleman's popular book *Emotional Intelligence* is a mixture of these two. Howard Gardner's well-known theory of 'multiple intelligences' clearly presents emotional intelligence as a separate type of intelligence. See Daniel Goleman, *Emotional Intelligence*, Bantam Doubleday: New York, 1996; Howard Gardner, *Frames of Mind*, Heinemann: New York, 1984.
3. Giovanna Colombetti and Evan Thompson, 'The feeling body: toward an enactive approach to emotion', in W. Overton, U. Muller and J. Newman (eds), *Body in Mind: Mind in Body: Developmental Perspectives on Embodiment and Consciousness*, Erlbaum: Mahwah, NJ, 2007.
4. This picture is largely agreed by the major figures in what is known as 'affective neuroscience'. See Antonio Damasio, *The Feeling of What Happens: Body, Emotion and the Making of Consciousness*, Vintage: New York, 2000; Joseph LeDoux, *Synaptic Self*, Viking: New York, 2002; Keith Oatley, *Best Laid Schemes*, Cambridge University Press: Cambridge, 1992; Jaak Panksepp, *Affective Neuroscience*, Oxford University Press: Oxford, 1998; Paul Ekman and Richard Davidson (eds), *The Nature of Emotion*, Oxford University Press: Oxford, 1995. For a recent review of the 'basic

emotions' argument, see Andrea Scarantino and Paul Griffiths, 'Don't give up on basic emotions', 2011, *Emotion Review*, 3(4), 444–454. Most of this research relies on people's judgement of the emotional expression on static images of faces. But in 'real life' we rely on dynamic changes – the information contained in changes of expression. A recent paper taking this into account finds evidence of only four basic emotional expressions: happy, sad, fearful/surprised and angry/disgusted. As these unfold over time, so further nuances and distinctions can be added. See Rachael Jack, Oliver Garrod and Philippe Schyns, 'Dynamic facial expressions of emotion transmit an evolving hierarchy of signals over time', *Current Biology*, 2014, 24, 187–192.

5. See for example Joel Benington and Craig Heller, 'Restoration of brain energy metabolism as the function of sleep', *Progress in Neurobiology*, 1995, 45, 347–360; Marcus Dworak et al., 'Sleep and brain energy levels: ATP changes during sleep', *Journal of Neuroscience*, 2010, 30(26), 9007–9016.

6. Silvan Tomkins, *Affect, Imagery, Consciousness*, Springer: New York, 2008.

7. Panksepp, op. cit., Chapter 8.

8. Gregory Bateson, Don D. Jackson, Jay Haley and John Weakland, 'Toward a theory of schizophrenia', *Behavioral Science*, 1956, 1(4), 251–254.

9. Mihaly Czikszentmihalyi, *Flow: The Psychology of Optimal Experience*, Harper & Row: London, 1990.

10. Barbara Frederickson, 'The role of positive emotions in positive psychology: the broaden-and-build theory of positive emotions', *American Psychologist*, 2001, 56, 218–226.

11. A good summary of this research is in Patricia Churchland, *Braintrust: What Neuroscience Tells Us about Morality*, Princeton University Press: Princeton, NJ, 2011.

12. Tom Peters, *Thriving on Chaos: Handbook for a Management Revolution*, Macmillan: London, 1989.

13. Elisabeth Kűbler-Ross, *On Death and Dying*, Scribner: New York, 1997.

14. D. Dutton and A. Aron, 'Some evidence for heightened sexual attraction under conditions of high anxiety', *Journal of Personality and Social Psychology*, 1974, 30, 510–517; D. Zillman, 'Attribution and misattribution of excitatory reaction', in J. Harvey, W. Ickes and R. Kidd (eds), *New Directions in Attribution Research, Vol. 2*, 1978, Erlbaum: Hillsdale, NJ.

15. Kristen Linquist, Tor Wager, Hedy Knober, Eliza Bliss-Moreau and Lisa Barrett, 'The brain basis of emotion: a meta-analytic review', *Behavioral and Brain Sciences*, 2012, 35(3), 121–143.

16. Keith Oatley and Jennifer Jenkins, *Understanding Emotions*, Blackwell: Oxford, 1996.

17. Mary Immordino-Yang, Andrea NcColl, Hanna Damasio and Antonio Damasio, 'Neural correlates of admiration and compassion', *Proceedings of the National Academy of Sciences*, 2009, 106 (19), 8021–8026.

18. Paul Griffiths and Andrea Scarantino, 'Emotions in the wild: the situated perspective on emotion', in Philip Robbins and Murat Aydede (eds), *The Cambridge Handbook of Situated Cognition*, Cambridge University Press: Cambridge, 2008.

19. Aristotle, *Nicomachean Ethics*, J.M. Dent: London, 1984.
20. Terrie Moffitt et al., 'A gradient of childhood self-control predicts health, wealth and social safety', 2011, www.pnas.org/cgi/doi/10.1073/pnas.1010076108.
21. Paula Niedenthal, Piotr Winkielman, Laurie Mondillon and Nicolas Vermeulen, 'Embodiment of emotion concepts', *Journal of Personality and Social Psychology*, 2009, 96(6), 1120-1136.
22. Harald Traue and James Pennebaker (eds), *Emotion Inhibition and Health*, Hogrefe and Huber: Seattle, 1993; James Pennebaker (ed.), *Emotion, Disclosure and Health*, American Psychological Association: Washington, DC, 1995. The negative effects of chronic muscular tension, designed to inhibit emotion, has sometimes been referred to as 'body armouring', a term coined by the controversial psychotherapist Wilhelm Reich. Whether Reich's reputation as a charlatan is deserved or not, his work on body armour has been largely vindicated by subsequent research.
23. Penelope Brown and Stephen Levinson, *Politeness: Some Universals in Language Usage*, Cambridge University Press: Cambridge, 1987.
24. Martina Ardizzi, Francesca Martini, Maria Umilta, Mariateresa Sestito, Roberto Ravera and Vittorio Gallese, 'When early experiences build a wall to others' emotions: an electrophysiological and autonomic study', PLoS ONE, 2013, 8(4): e61004. doi:10.1371/journal.pone.0061004.
25. David Neal and Tanya Chartrand, 'Embodied emotion perception: amplifying and dampening facial feedback modulates emotional perception accuracy', *Social Cognition and Personality Science*, 2011, 2(6), 673-678. Andreas Hennenlotter, Christian Dresel, Florian Castrop, Andres Baumann, Afra Wohlschläger and Bernhard Haslinger, 'The link between facial feedback and neural activity within central circuits of emotion: new insights from botulinum toxin-induced denervation of frown muscles', *Cerebral Cortex*, 2009, 19, 537-542.
26. Paula Niedenthal, 'Embodying emotion', *Science*, 2007, 316, 1002-1005.
27. Jane Richards, 'The cognitive consequences of concealing feelings', *Current Directions in Psychological Science*, 2004, 13(4), 131-134.
28. See Roy Baumeister and John Tierney, *Willpower: Rediscovering Our Greatest Strength*, Allen Lane: London, 2011. For the 'gargling' studies, see M. Hagger and N. Chatzisarantis, 'The sweet taste of success: the presence of glucose in the oral cavity moderates the depletion of self-control resources', *Personality and Social Psychology Bulletin*, 2012, doi: 10.1177/0146167212459912; and M. Sanders, S. Shirk, C. Burgin and L. Martin, 'The gargle effect: rinsing the mouth with glucose enhances self-control', *Psychological Science*, 2012, doi: 10.1177/0956797612450034.
29. Michael Trimble, *Why Humans Like to Cry: Tragedy, Evolution and the Brain*, Oxford University Press: Oxford, 2012.
30. This analysis is similar to that of William Frey, *Crying: The Mystery of Tears*, Winston Press: Minneapolis, MN, 1985.
31. Shani Gelstein, Yaara Yeshurun, Liron Rozenkrantz, Sagit Shushan, Idan Frumin, Yehudah Roth and Noam Sobel, 'Human tears contain a chemosignal', *Science*, 6 January 2011, doi: 10.1126/science.1198331. Ad

Vingerhoets, *Why Only Humans Weep*, Oxford University Press: Oxford, 2013.

32. Jonathan Haidt, *The Righteous Mind*, Allen Lane: London, 2012.

33. For a review of this research, see Joseph Forgas and Eric Eich, 'Affective influences on cognition', in A. Healy and R. Proctor (eds), *Handbook of Psychology, Vol. 4: Experimental Psychology*, Wiley: New York, 2014.

34. Isabelle Blanchette and Serge Caparos, 'When emotions improve reasoning: the possible roles of relevance and utility', *Thinking and Reasoning*, 2013, 19(3), 399–413.

35. Eveline Crone, Riek Somsen, Bert van Beek and Mauritz van der Molem, 'Heart rate and skin conductance analysis of antecedents and consequences of decision-making', *Psychophysiology*, 2004, 41, 531–540.

36. Natalie Werner, Katharina Jung, Stefan Duschek and Rainer Schandry, 'Enhanced cardiac perception is associated with benefits in decision-making', *Psychophysiology*, 2009, 46, 1123–1129.

Chapter 7

1. The general picture I am going to sketch here draws on a number of similar approaches: Antonio Damasio (2010), op. cit.; Michael Arbib, Brad Gasser and Victor Barrès, 'Language is handy but is it embodied?', *Neuropsychologia*, 2014, 55, 57–70; Friedemann Pulvermüller and Luciano Fadiga, 'Active perception: sensorimotor circuits as a cortical basis for language', *Nature Reviews Neuroscience*, 2010, 11, 351–360; Rolf Zwaan, 'Embodiment and language comprehension: reframing the discussion', *Trends in Cognitive Sciences*, 2014, 18(5), 229–234; Arthur Glenberg and Vittorio Gallese, 'Action-based language: a theory of language acquisition, comprehension and production', *Cortex*, 2012, 48(7), 905–922.

2. Michael Masson, Daniel Bub and Meaghan Newton-Taylor, 'Language-based access to gestural components of conceptual knowledge', *Quarterly Journal of Experimental Psychology*, 2008, 61(6), 869–882.

3. For an excellent account of how financial dealers operate, and the hormonal influences behind their decisions, see John Coates, *The Hour between Dog and Wolf: Risk Taking, Gut Feelings and the Biology of Boom and Bust*, Penguin: London, 2012.

4. There is a most beautiful fjord in the south-west corner of New Zealand called Doubtful Sound. I'm sure you could make the appropriate noise . . .

5. Johnson, op. cit, in a chapter cheekily (*sic*) entitled 'Feeling William James's "But"'. The embodied feeling of doubt is given a celebrated discussion in James's magisterial *Principles of Psychology*, Dover: New York, 1890/1950.

6. Geoffrey Willans and Ronald Searle, *How to Be Topp*, Max Parrish: London, 1955.

7. See Gabriella Vigliocco, Stavroula-Thaleia Kousta, Pasquale della Rosa, David Vinson, Marco Tettamanti, Joseph Devlin and Stefano Cappa, 'The neural representation of abstract words: the role of emotion', *Cerebral Cortex*, 13 February 2013, doi:10.1093/cercor/bht025.

8. William James, *Some Problems of Philosophy*, Harvard University Press: Cambridge, MA, 1911/1979, 32–33.
9. Lakoff and Johnson, op. cit., pp. 43, 17.
10. Michael Masson et al., op. cit.
11. John Bargh, Mark Chen and Lara Burrows, 'Automaticity of social behaviour: direct effects of trait construct and stereotype activation on action', *Journal of Personality and Social Psychology*, 1996, 71 (2), 230–244. We should note that there have been a number of failures to replicate this study, but John Bargh has stoutly defended the validity of his experiments.
12. Arthur Glenberg, Marc Sato, Luigi Cattaneo, Lucia Riggio, Daniele Palumbo and Giovanni Buccino, 'Processing abstract language modulates motor system activity', *Quarterly Journal of Experimental Psychology*, 2008, http://dx.doi.org/10.1080/17470210701625550; and Veronique Boulenger, Olaf Hauk and Friedemann Pulvermüller, 'Grasping ideas with the motor system: semantic somatotopy in idiom comprehension', *Cerebral Cortex,* 2009, 19, 1905–1914.
13. Glenberg et al., op. cit., p.13.
14. Pulvermüller and Fadiga, op. cit.
15. U. Tan, M. Okuyan and A. Akgun, 'Sex differences in verbal and spatial ability reconsidered in relation to body size, lung volume and sex hormones', *Journal of Perceptual and Motor Skills*, 1996, 3 (2), 1347–1360.
16. Mirjam Tuk, Debra Trampe and Luk Warlop, 'Inhibitory spillover: increased urination urgency facilitates impulse control in unrelated domains', *Psychological Science*, 2011, 22, 627–633.
17. Dana Carney, Amy Cuddy and Andy Yap, 'Power posing: brief non-verbal displays affect neuro-endocrine levels and risk tolerance', *Psychological Science*, 2010, 21 (10), 1363–1368. See also John Coates, op. cit. Some of the examples in this section are gratefully borrowed from Mark Johnson (2007) op. cit.
18. Pablo Brinol and Richard Petty, 'Embodied persuasion: fundamental processes by which bodily processes can impact attitudes', in Gün Semin and Eliot Smith (eds), *Embodiment Grounding: Social, Cognitive, Affective and Neuroscientific Approaches*, Cambridge University Press: Cambridge, 2008.
19. Lawrence Williams and John Bargh, 'Experiencing physical warmth promotes interpersonal warmth', *Science*, 2008, 322, 606–607.
20. Harry Harlow and Robert Zimmermann, 'The development of affective responsiveness in infant monkeys', *Proceedings of the American Philosophical Society*, 1958, 102, 501–509.
21. Here I am drawing heavily on the work of George Lakoff and Rafael Núñez, *Where Mathematics Comes From: How the Embodied Mind Brings Mathematics into Being*, Basic Books: New York, 2000. And also on Paul Lockhart, *A Mathematician's Lament*, http://www.maa.org/sites/default/files/pdf/devlin/LockhartsLament.pdf
22. This is a very condensed discussion of an argument in Rafael Núñez, Laurie Edwards and João Matos, 'Embodied cognition as grounding for situatedness and context in mathematics education', *Educational Studies in Mathematics*, 1999, 39, 45–65.

23. G. Yue and K. Cole, 'Strength increases from the motor program: comparison of training with maximal voluntary and imagined muscle contractions', *Journal of Neurophysiology*, 1992, 67, 1114–1123.

24. Jean Decety and Jennifer Stevens, 'Action representation and its role in social interaction', in Keith Markman et al. (eds), *Handbook of Imagination and Mental Simulation*, Psychology Press: New York, 2009.

25. Jacques Hadamard, *The Psychology of Invention in the Mathematical Field*, Princeton University Press: Princeton, NJ, 1945, pp. 142–143.

26. John Allpress, personal communication; Gillian Lynne, choreographer of *Cats*, as told to Ken Robinson in *The Element: How Finding Your Passion Changes Everything*, Allen Lane: London, 2009.

27. Daniel Aiello, Andrew Jarosz, Patrick Cushen and Jennifer Wiley, 'Firing the executive: when an analytic approach does and does not help problem solving', *The Journal of Problem Solving*, 2012, 4(2), 116–127. The answer is 'figure', by the way.

28. Don Tucker, *Mind from Body: Experience from Neural Structure*, Oxford University Press: Oxford, 2007, p. 59.

Chapter 8

1. Parts of this chapter are based on Guy Claxton, 'How conscious experience comes about, and why meditation is helpful', in Michael West (ed.), *The Psychology of Meditation, 2nd edition*, Oxford University Press: Oxford, 2015.

2. Alan Baddeley, *Working Memory, Thought and Action*, Oxford University Press: Oxford, 2007.

3. Bernard Barrs, 'Global workspace theory of consciousness: toward a cognitive neuroscience of human experience', *Progress in Brain Research*, 2005, 150, 45–53.

4. Benny Shanon, 'Against the spotlight model of consciousness', *New Ideas in Psychology*, 2001, 19, 77–84.

5. Heinz Werner, 'Microgenesis and aphasia', *Journal of Abnormal Social Psychology*, 1956, 52, 347–353.

6. Jason Brown, Professor of Neurology at New York University, has developed the idea of microgenesis further, and explored its similarity to images of consciousness found in various traditional Buddhist teachings such as the Abhidharma. See Jason Brown, 'Microgenesis and Buddhism: the concept of momentariness', *Philosophy East and West*, 1999, 49(3), 261–277; also by the same author, *Self and Process: Brain States and the Conscious Present*, Springer-Verlag: New York, 1991.

7. Shanon, op. cit., p. 81.

8. Susan Goldin-Meadow and Susan Wagner, 'How our hands help us learn', *Trends in Cognitive Science*, 9(5), 2005, 234–241; Andy Clark, *Supersizing the Mind*, Oxford University Press: Oxford, 2008.

9. Traditionally, Native American languages such as Hopi do not require this insertion of self. See John Carroll (ed.), *Language, Thought and Reality: Selected Writings of Benjamin Lee Whorf*, MIT Press: Boston, MA, 1956.

10. See, for a Western philosophical version of this argument, Derek Parfit, *Reasons and Persons*, Oxford University Press: Oxford, 1986; or, for a Buddhist treatment of the same argument, David Kalupahana, *The Principles of Buddhist Psychology*, SUNY Press: Albany, NY, 1987. This argument is also familiar from the schools of 'general semantics' of Alfred Korzybski, *Science and Sanity*, Institute of General Semantics: Brooklyn, NY, 1944.

11. Readers who know about Cognitive Behavioural Therapy, or contemporary interpretations of Buddhism, will know that I have skimmed over some very complex territory here. For more detail, try Mark Williams, John Teasdale, Zindel Segal and Jon Kabat-Zinn, *The Mindful Way through Depression*, Guilford Press: London, 2007; and Stephen Batchelor, *Buddhism without Beliefs*, Riverhead Press: New York, 1997.

12. One such story has the French mathematician Henri Poincaré putting aside the problem he had been wrestling with to go on a bus trip with friends. He recalled, 'At the moment when I put my foot on the [bus] step, the idea came to me, without anything in my former thoughts seeming to have paved the way for it, that the transformations I had used to define the Fuchsian functions were identical with those of non-Euclidean geometry.' See Jacques Hadamard, *An Essay on the Psychology of Invention in the Mathematical Field*, Dover Publications: New York, 1954.

13. Andy Clark, 'Whatever next? Predictive brains, situated agents and the future of cognitive science', *Behavioral and Brain Sciences*, 2013, 36(3), 181–204.

14. T. L. Friedman, 'How to get a job at Google', *New York Times*, 22 February 2014.

15. L. Carmichael, H. P. Hogan and A. A. Walter, 'An experimental study of the effect of language on the reproduction of visually perceived form', *Journal of Experimental Psychology*, 1932, 15, 73–86.

16. Colin Martindale, 'Creativity and connectionism', in S. Smith, T. Ward and R. Finke (eds), *The Creative Cognition Approach*, MIT Press: Cambridge, MA, 1995.

17. Stephen Krashen, *Principles and Practice in Second Language Acquisition*, Pergamon Press: Oxford, 1982.

18. Heather Stewart, 'This is how we let the credit crunch happen, Ma'am', *The Guardian*, 26 July 2009.

19. Peter Fensham and Ference Marton, 'What has happened to intuition in science education?', *Research in Science Education*, 1992, 22, 114–122; Jacques Maritain, *Creative Intuition in Art and Poetry*, Princeton University Press: Princeton NJ, 1977.

20. Marcel Kinsbourne, 'What qualifies a representation for a role in consciousness?', in Jonathan Cohen and Jonathan Schooler (eds), *Scientific Approaches to Consciousness*, Erlbaum: Mahwah NJ, 1997, 335–355.

21. Benjamin Libet, 'Brain stimulation in the study of neuronal functions for conscious sensory experience', *Human Neurobiology*, 1982, 1, 235–242.

22. S. Johnson, J. Marro and J. J. Torres, 'Robust short-term memory without synaptic learning', *PLoS ONE*, 2013, 8 (1), e50276, doi:10.1371/journal.pone.0050276.

23. Gerald Edelman and Giulio Tononi, *Consciousness: How Matter Becomes Imagination*, Allen Lane: London, 2000; also Damasio (2010), op. cit., and Llinás, op. cit.

24. Justin Hepler and Dolores Albarracin, 'Complete unconscious control: using (in)action primes to demonstrate completely unconscious activation of inhibitory control mechanisms', *Cognition*, 2013, 128(3), 271–279.

25. Adrian Wells, *Metacognitive Therapy for Anxiety and Depression*, Guilford Press: London, 2008.

26. Guy Claxton, 'Mindfulness, learning and the brain', *Journal of Rational-Emotive and Cognitive-Behavioral Therapy*, 2006, 23(4), 301–314.

27. Peter Carruthers, 'How we know our own minds: the relationship between mind-reading and metacognition', *Behavioral and Brain Sciences*, 2009, 32(2), 121–138.

28. Carina Remmers, Sandra Topolinski and Johannes Michalak, 'Mindful intuition: does mindfulness influence the access to intuitive processes?', *Journal of Positive Psychology*, 2014, doi: 10.1080/17439760.2014.950179, emphasis added.

29. Timothy Wilson, *Strangers to Ourselves: Discovering the Adaptive Unconscious*, Belknap Press: Cambridge, MA, 2002.

Chapter 9

1. Blakeslee and Blakeslee, op. cit., p. 3. I am drawing heavily on this great book in the first part of this chapter.

2. The peripersonal space relates closely to the 'egocentric perspective' that we discussed in Chapter 5.

3. Michael Graziano and Charles Gross, 'Vision, movement and the monkey brain', in A. Mikami (ed.), *The Association Cortex: Structure and Function*, Amsterdam: Overseas Publishers Association, 1997.

4. Richard Abrams, Christopher Davoli, Feng Du, William Knapp and Daniel Paull, 'Altered vision near the hands', *Cognition*, 2008, 107(3), 1035–1047.

5. Anna Berti and Francesca Frassinetti, 'When far becomes near: re-mapping of space by tool-use', *Trends in Neurosciences*, 2000, 20, 560–564.

6. Berenice Valdés-Conroy, Francisco Román, Jose Hinojosa and Paul Shorkey, 'So far so good: emotion in the peripersonal/extrapersonal space', *PLOS One*, 2012, 7(11), e49162.

7. Angelo Maravita and Atsushi Iriki, 'Tools for the body (schema)', *Trends in Cognitive Science*, 2004, 8(2), 79–86; Lucilla Cardinali, Francesca Frassinetti, Claudio Brozzoli, Christian Urquizar, Alice Roy and Alessandro Farné, 'Tool-use induces morphological updating of the body schema', *Current Biology*, 2009, 19(12), 478–479.

8. Blakeslee and Blakeslee, op. cit., p. 151.

9. Ibid., p. 206. For a good up-to-date review of this work, see Henrik Ehrsson, 'The concept of body ownership and its relation to multisensory integration', in B. Stein (ed.), *The New Handbook of Multisensory Processes*,

MIT Press: Cambridge, MA, 2012. Watch the rubber hand illusion at http://www.youtube.com/watch?v=sxwn1w7MJvk

10. Nick Yee and Jeremy Bailenson, 'The difference between being and seeing: the relative contribution of self-perception and priming to behavioral changes via digital self-representation', *Media Psychology*, 2009, 12, 195–209.

11. Blakeslee and Blakeslee, op. cit., p. 147.

12. Francis Evans, 'Two legs, think using and talking: the origins of the creative engineering mind', unpublished manuscript, available at http://www.timhunkin.com/a119_francis_evans.htm.

13. I have borrowed the phrase 'person plus' from David Perkins, 'Person plus: a distributed view of thinking and learning', in Gavriel Salomon (ed.), *Distributed Cognitions: Psychological and Educational Considerations*, Cambridge University Press: Cambridge, 1996.

14. Richard Dawkins, *The Extended Phenotype*, Oxford University Press: Oxford, 1982.

15. These examples, like much of the thinking in this section (and indeed the whole spirit of the book), draws on the massively exciting and elegant writing of Scottish philosopher Andy Clark. See especially his books *Being There, Natural-Born Cyborgs, and Supersizing the Mind*.

16. See the discussion in Clark (2008), op. cit. And, for example, Fred Adams and Ken Aizawa, 'Why the mind is still in the head', in M. Aydede and P. Robbins (eds), *Cambridge Handbook of Situated Cognition*, Cambridge University Press: Cambridge, 2008.

17. Steven Mithen, 'Comments of natural born cyborgs by Andy Clark', *Metascience*, 2004, 13(7), 163–169.

18. There is an old puzzle about a man who wakes up one morning, looks at a stick by the side of his bed, lets out a howl of despair, and kills himself. Outside the room, another man, hearing the shot, laughs silently to himself. By questioning, you have to find out the occupation of this second man. Eventually, if you can be bothered to persist, you will discover that this man is the Second Smallest Man In The World, and in the night he has sawn a small piece off the end of the measuring stick that the Smallest Man In The World uses every morning to make sure that he still deserves the title ... By altering the measuring stick, this jealous dwarf has become a cunning (if implausible) assassin.

19. Vittorio Gallese, 'Embodied simulation: from neurons to phenomenal experience', *Phenomenology and the Cognitive Sciences*, 2005, 4, 23–48. A great deal has been written about mirror neurons in the popular neuroscience literature in the last ten years, so I make this discussion very brief. There are still those who are unconvinced that the research on monkeys has been adequately replicated in human beings, but the bulk of opinion now accepts the view I am expressing here.

20. Yuri Hasson, Nuval Nir, Ifat Levy, Galit Fuhrmann and Raphael Malach, 'Intersubject synchronization of cortical activity during natural vision', *Science*, 2004, 303(12), 1634–1640; Michael Richardson, Kerry Marsh, Robert Isenhower, Justin Goodman and Richard Schmidt, 'Rocking

together: dynamics of intentional and unintentional interpersonal coordination', *Human Movement Science*, 2007, 26, 867–891.

21. Greg Stephens, Lauren Silbert and Uri Hasson, 'Speaker–listener neural coupling underlies successful communication', *PNAS*, 2010, 107(32), 14425–14450.

22. Fabian Ramseyer, 'Non-verbal synchrony in psychotherapy: embodiment at the level of the dyad', in Wolfgang Tschacher and Claudia Bergomi (eds), *The Implications of Embodiment: Cognition and Communication*, Imprint Academic: Exeter, 2011.

23. Jody Osborn and Stuart Derbyshire, 'Pain sensation evoked by observing injury in others', *Pain*, 2010, 148, 268–274.

24. Mark Williams, Adam Morris, Frances McGlone, David Abbott and Jason Mattingley, 'Amygdala responses to fearful and happy facial expressions under conditions of binocular suppression', *Journal of Neuroscience*, 2004, 24(12), 2898–2904.

25. Richard Wiseman and Caroline Watt, 'Judging a book by its cover: the unconscious influence of pupil size on consumer choice', *Perception*, 2010, 39, 1417–1419.

26. Sun Tzu, *The Art of War*, Penguin: London, 2008 edn.

27. Nicholas Humphrey, 'The social function of intellect', in Patrick Bateson and Robert Hinde (eds), *Growing Points in Ethology*, Cambridge University Press: Cambridge, 1976.

28. Lauren Adamson, Heidelise Als, Edward Tronick and Berry Brazelton, 'The development of social reciprocity between a sighted infant and her blind parents: a case study', *Journal of the American Academy of Child Psychiatry*, 1977, 16(2), 194–207.

29. Lynne Murray and Peter Cooper, 'Intergenerational transmission of affective and cognitive processes associated with depression: infancy and the preschool years', in I. Goodyer (ed.), *Unipolar Depression: A Lifespan Perspective*, Oxford University Press: Oxford, 2003.

30. Thomas Fuchs, 'Depression, intercorporeality and interaffectivity', *Journal of Consciousness Studies*, 2013, 20(7–8), 219–238.

31. See Colombetti and Thompson, op. cit. I am also drawing here on Kinsbourne and Jordan, op. cit.

32. Chris Frith, 'Consciousness is for other people', *Behavioral and Brain Sciences*, 1995, 18, 682–683.

33. Bruce MacLennan, 'Evolutionary neuro-theology and the varieties of religious experience', in R. Joseph (ed.), *Neuro-theology: Brain, Science, Spirituality, Religious Experience*, University Press: San Francisco, CA, 2002; Jean Knox, *Self-Agency in Psychotherapy*, W. W. Norton: New York, 2011.

Chapter 10

1. Andrew Harrison, *Making and Thinking: A Study of Intelligent Activities*, Hassocks: Harvester, 1978, p. 1.

2. Francis Fukuyama, 'Making things work: review of Matthew Crawford's *Shop Class as Soul Craft*', *New York Times*, 7 June 2009.

3. Mike Rose, *The Mind at Work: Valuing the Intelligence of the American Worker*, Allen Lane: New York, 2004; also Mike Rose, 'The working life of a waitress', *Mind, Culture and Activity*, 2001, 8(1), 3–27.
4. Quoted in Rose (2001), op. cit.
5. See Keith Markham, William Klein and Julie Suhr (eds), *Handbook of Imaginational and Mental Simulation*, Psychology Press: Hove, 2009.
6. Rose (2004), op. cit., pp. 100-101.
7. Jeanne Bamberger, 'The laboratory for making things: developing multiple representations of knowledge', in D. A. Schon (ed.), *The Reflective Turn: Case Studies in and on Educational Practice*, Teachers College Press: New York, 1991.
8. Michele Root-Bernstein and Robert Root-Bernstein, 'Thinkering', *The Creativity Post*, 18 February 2012: http://www.creativitypost.com/psychology/thinkering. See also their book *Sparks of Genius*, Houghton Mifflin: Boston, 1999.
9. Chen-Bo Zhong, Ap Dijksterhuis and Adam Galinsky, 'The merits of unconscious thought in creativity', *Psychological Science*, 2008, 19(9), 912–918; Kristin Flegal and Michael Anderson, 'Overthinking skilled motor performance: or why those who teach can't do', *Psychonomic Bulletin and Review*, 2008, 15(5), 927–932.
10. Dianne Berry and Donald Broadbent, 'On the relationship between task performance and associated verbalizable knowledge', *Quarterly Journal of Experimental Psychology*, 1984, 36(A), 209–231.
11. Hubert and Stuart Dreyfus, *Mind over Machine: The Power of Human Intuition and Expertise in the Era of the Computer*, Free Press: New York, 1986.
12. Michael Polanyi, *Personal Knowledge: Towards a Post-Critical Philosophy*, Routledge and Kegan Paul: London, 1958.
13. In e.e. cummings, *A Miscellany*, Peter Owen: London, 1965.
14. In one of Dorothy L. Sayers' Lord Peter Wimsey novels, *Unnatural Death*, a former Lyons Corner House 'nippy' congratulates Wimsey on his observation skills by saying 'You're a noticing one, aren't you? ... Make a good waiter you would – not meaning any offence, sir, that's a real compliment from one who knows.'
15. Quotes from Root-Bernstein and Root-Bernstein (1999) op. cit. I'm drawing liberally from their marvellous book in these paragraphs.
16. Barbara Rogoff, *The Cultural Nature of Human Development*, Oxford University Press: New York, 2003.
17. Richard Sennett, *The Craftsman*, Allen Lane: London, 2008.
18. Buster Olney, 'Speedy feet, but an even quicker thinker', *New York Times*, 1 February 2002.
19. Philippa Lally, Cornelia van Jaarsfeld, Henry Potts and Jane Wardle, 'How are habits formed? Modelling habit formation in the real world', *European Journal of Social Psychology*, 2010, 40(6), 998–1009.
20. See Harald Jorgensen and Susan Hallam, 'Practising', in S. Hallam, I. Cross and M. Thaut (eds), *Oxford Handbook of the Psychology of Music*, Oxford University Press: Oxford, 2008.

21. Sennett (2008), op. cit.
22. Seymour Papert and Idit Harel, *Constructionism*, Ablex Publishing, Norwood, 1991.
23. Rose (2004), op. cit., p. xxxii.

Chapter 11

1. William James, *Principles of Psychology, Vol. 1*, Dover: New York, 1890, p. 696.
2. Stephen Porges, Body Awareness Questionnaire, 1993, available online at www.stephenporges.com.
3. Wolf Mehlings, Cynthia Price, Jennifer Daubenmier, Mike Acree and Elizabeth Bartmess, 'The multidimensional assessment of interoceptive awareness', *PLOS ONE*, 2012, 7(11), e48230.
4. Daniella Furman, Christian Waugh, Kalpa Bhattacharjee and Renee Thompson, 'Interoceptive awareness, positive affect and decision making in Major Depressive Disorders', *Journal of Affective Disorders*, 2013, dx.doi.org/10.1016/j.jad.2013.06.044.
5. Sahib Khalsa, David Rudrauf, Antonio Damasio, Richard Davidson, Antoine Lutz and Daniel Tranel, 'Interoceptive awareness in experienced meditators', *Psychophysiology*, 2008, 45(4), 671–678; Kieran Fox, Pierre Zakarauskas, Matt Dixon, Melissa Ellamil and Evan Thompson, 'Meditation experience predicts introspective accuracy', *PLOS ONE*, 2012, 7(9), e45370.
6. Bryan Lask, Isky Gordon, Deborah Christie and Ian Frampton, Uttom Chowdhury and Beth Watkins, 'Functional neuroimaging in early-onset anorexia nervosa', *International Journal of Eating Disorders*, 2005, 37, 49–51.
7. These studies are discussed more fully in my *Hare Brain, Tortoise Mind: Why Intelligence Increases When You Think Less*, Fourth Estate: London, 1997. See also Ap Dijksterhuis, 'Think different: the merits of unconscious thoughts in preference development and decision-making', *Journal of Personality and Social Psychology*, 2004, 87(5), 586–598.
8. Carolyn Yucha and Doil Montgomery, 'Evidence-Based Practice in Biofeedback and Neurofeedback', *Applied Psychophysiology and Biofeedback*, 2008. The studies citied in this section are cited in this review unless otherwise referenced.
9. Maman Paul, Kanupriya Garg and Jaspal Singh Sandhu, 'Role of biofeedback in optimising psychomotor performance in sports', *Asian Journal of Sports Medicine*, 2012, 3(1), 1–7.
10. Kieran Cox et al., 'Meditation experience predicts introspective accuracy', *PLOS ONE*, 2012, 7(9), e45370; Wendy Hasenkamp and Lawrence Barsalou, 'Effect of meditation experience on functional connectivity of distributed brain networks', *Frontiers in Human Neuroscience*, 2012, 6, 1–14. But for a more complex view see Khalsa et al., op. cit.
11. Norman Farb, Zindel Segal, Helen Mayberg, Jim Bean, Deborah McKeon, Zainab Fatima and Adam Anderson, 'Attending to the present: mindfulness meditation reveals distinct neural modes of self-reference', *Social Cognitive and Affective Neuroscience*, 2007, 2(4), 313–322.

12. Richard Davidson et al., 'Alteration in brain and immune function produced by mindfulness meditation', *Psychosomatic Medicine*, 2003, 65(4), 564–570.

13. Amishi Jha, Jason Krompinger and Michael Baime, 'Mindfulness training modifies subsystems of attention', *Social Cognitive and Affective Neuroscience*, 2007, 7, 101–119.

14. Heleen Slagter, Antoine Lutz, Lawrence Greischar, Andrew Francis, Sander Nieuwenhuis, James Davis and Richard Davidson, 'Mental training affects distribution of limited brain resource's', *PLOS Biology*, 2007, e138.

15. Ulrich Kirk, Jonathan Downar and Read Montague, 'Interoception drives increasingly rational decision-making in meditators playing the ultimate game', *Frontiers in Neuroscience*, 2011, doi: 10.3389/fnins.2011. 00049.

16. Quoted in Wolf Mehling et al., 'Body awareness: a phenomenological inquiry into the common ground of mind-body therapies', *Philosophy, Ethics and Humanities in Medicine*, 2011, 6(6), http://www.peh-med. com/content/6/1/6.

17. Eugene Gendlin and Joseph Rychlak, 'Psychotherapeutic processes', *Annual Review of Psychology*, 2000, 21, 155–190. See also Jean Knox's excellent review of this work in *Self-Agency in Psychotherapy: Attachment, Autonomy and Intimacy*, W. W. Norton: New York, 2011.

18. Eugene Gendlin, *Focusing: How to Gain Direct Access to your Body's Knowledge*, Rider: London, 2003.

19. Eugene Gendlin, *Sitting with Gene at his Leading Edge*, telephone course, March/April 2011, Focusing Resources: Berkeley, CA.

20. Louis Cozolino, *The Neuroscience of Psychotherapy*, W. W. Norton: New York, 2010. A speculative but well-informed paper is by Peter Afford, 'Focusing in an age of neuroscience', *The Folio*, 2012, 23(1), 66–83. There is also a wealth of information about other kinds of therapy that give a central role to the body in Courtenay Young (ed.), *About the Science of Body Psychotherapy*, Body Psychotherapy Publications: Galashiels, Scotland, 2012.

21. See Klaus Meyerson, 'Thinking at the edge in industry', Mary Larrabee, 'Eighth graders think at the edge', and Greg Walkenden, 'Thinking at the edge in environmental management', all in *The Folio*, 2004, 19(1), 94–111.

22. E.g. Rod Dishman and Patrick O'Connor, 'Lessons in exercise neurobiology: the case of endorphins', *Mental Health and Physical Activity*, 2009, 2, 4–9.

23. Henriette van Praag, 'Neurogenesis and exercise: past and future directions', *Neuromolecular Medicine*, 2008, 10, 128–140; Bernard Winter et al., 'High impact running improves learning', *Neurobiology of Learning and Memory*, 2007, 87, 597–609; Michelle Voss et al., 'Functional connectivity: a source of variance in the association between cardiorespiratory fitness and cognition?', *Neuropsychologia*, 2010, 48, 1394–1406.

24. Joseph Donnelly et al., 'Physical activity across the curriculum (PAAC): a randomised control trial to promote physical activity and diminish overweight and obesity in elementary school children', *Preventative Medicine*,

2009, 49, 336–341; Yu-Kai Chang, Yu-Jung Tsai, Tai-Ting Chen and Tsung-Min Hung, 'The impact of coordinative exercise on executive function in kindergarten children: an ERP study', *Experimental Brain Research*, 2013, 225, 187–196.

25. Thomas Fuchs and Sabine Koch, 'Embodied affectivity: on moving and being moved', *Frontiers in Psychology*, 2014, 5, article 508.

26. Peter Wayne et al., 'Effects of tai chi on cognitive performance in older adults: systematic review and meta-analysis', *Journal of the American Geriatric Society*, 2014, 62(1), 25–39.

27. Neha Gothe et al., 'The effect of acute yoga on executive function', *Journal of Physical Activity and Health*, 2013, 10(4), 488–495.

28. Peter Lovatt, 'Dance psychology', *Psychology Review*, September 2013, 18–21; Carine Lewis and Peter Lovatt, 'Breaking away from set patterns of thinking: improvisation and divergent thinking', *Thinking Skills and Creativity*, 2013, 9, 46–58.

Chapter 12

1. Clark (1997), op. cit., pp. 220–221.

2. I have just discovered a great paper that explores this interpersonal resonance in detail: Tom Froese and Thomas Fuchs, 'The extended body: a case study in the neurophenomenology of social interaction', *Phenomenology and the Cognitive Sciences*, 2012, 11, 205–235.

3. I use the word 'get' rather than 'understand' deliberately. Getting a joke is a deeply visceral event. Understanding engages what I have called the Word-Scape, but often nothing more. Just to *understand* embodiment seems a bit of a waste.

4. Sir Christopher Frayling, interviewed in *The Guardian*, 29 June 2004.

5. I and my colleagues at the Centre for Real-World Learning at the University of Winchester have written extensively about this. See, for example, Guy Claxton and Bill Lucas, 'Is vocational education for the less able?', in Philip Adey and Justin Dillon (eds), *Bad Education: Debunking Myths in Education*, McGraw-Hill: London, 2012.

6. Nick Duffell, *The Making of Them: The British Attitude to Children and the Boarding School System*, Lone Arrow Press: London, 2000; Nick Duffell, *Wounded Leaders: British Elitism and the Entitlement Illusion*, Lone Arrow Press: London, 2014; Nick Duffell, 'Why boarding schools produce bad leaders', *The Guardian*, 9 June 2014.

7. http://www.theguardian.com/theobserver/2014/may/10/the-big-issue-boarding-schools-abuse.

8. Matthew Crawford, *The Case for Working with Your Hands*, Allen Lane: London, 2009, p. 3.

9. Martin Duerden, Tony Avery and Rupert Payne, *Polypharmacy and Medicines Optimisation*, The King's Fund: London, 2013.

10. A. Prentice and T. Lind, 'Fetal heart rate monitoring during labour – too frequent intervention, too little benefit', *The Lancet*, 1987, 330(8572), 1375–1377. For a review of the cost of encouraging midwives to ignore their intuition, see Robbie Davis-Floyd and Elizabeth Davis, 'Intuition as

authoritative knowledge in midwifery and homebirth', in Robbie Davis-Floyd and Sven Arvidson (eds), *Intuition: Interdisciplinary Perspectives*, Routledge: London, 1997.

11. See, for example, John Kirkland, *Crying and Babies*, Croom Helm: London, 1985.

12. Sara Konrath, Edward O'Brien and Courtney Hsing, 'Changes in dispositional empathy in American college students over time: a meta-analysis', *Personality and Social Psychology Review*, 2014, 40, 1079–1091.

13. Quoted in Nicholas Carr, *The Shallows: How the Internet is Changing the Way We Think, Read and Remember*, W. W. Norton: New York, 2010.

14. Mark Bauerlein, 'Why Johnny Gen-X can't read nonverbal cues', *Wall Street Journal*, 4 September 2009.

15. Carr, op. cit.

16. Pascal Zachary, 'Digital designers rediscover their hands', *New York Times*, 14 September 2008.

17. I have learned much that underpins this discussion from Matthew Crawford, *Shop Class as Soul Craft: An Inquiry into the Value of Work*, Allen Lane: New York, 2009.

18. Richard Sennett, *The Corrosion of Character: The Personal Consequences of Work in the New Capitalism*, W. W. Norton: New York, 1998.

19. Matthew Crawford, 'The case for working with your hands', *New York Times*, 24 May 2009.

20. Personal communication from Art Costa. For his latest thinking, see Art Costa and Bena Kallick, *Dispositions: Reframing Teaching and Learning*, Corwin: San Francisco, 2014.

21. Tom de Castella, 'Is working with your hands better than just with your head?', *BBC News Magazine*, 4 January, 2011.

22. Rolf Pfeifer, quoted in the *Boston Globe*, 13 January 2008.

23. Stephen Ceci, *On Intelligence*, Harvard University Press: Cambridge, MA, 1996; Michael Howe, *IQ in Question*, Sage: London, 1997; David Perkins, *Outsmarting IQ*, Free Press: New York, 1995; Ken Richardson, *The Making of Intelligence*, Weidenfeld and Nicolson: London, 1999.

24. Stephen Ceci and Jeffrey Liker, 'A day at the races: a study of IQ, expertise and cognitive complexity', *Journal of Experimental Psychology: General*, 1986, 115, 255–266; Sylvia Scribner, 'Studying working intelligence', in B. Rogoff and J. Lave (eds), *Everyday Cognition: Its Development in Social Context*, Harvard University Press: Cambridge, MA, 1984, quote p. 26.

25. G. Mugny and F. Carugati (eds), *Social Representation of Intelligence*, Pergamon: Oxford, 1989.

26. Laszlo Bock, quoted in *The Guardian*, 24 February 2104 and the *New York Times*, 22 February 2014; Keith Stanovich and Richard West, 'What intelligence tests miss', *The Psychologist*, 2014, 27(2), 80–83; also Keith Stanovich, *What Intelligence Tests Miss: The Psychology of Rational Thought*, Yale University Press: New Haven, CT, 2009. Stanovich has felt the need to invent a new kind of mental disorder, which he called *dysrationalia*: the inability to think and behave rationally (in one's own best interests, say), despite performing perfectly respectably on IQ tests!

27. Psychology as a discipline still suffers from two disorders we could call Polarisation Syndrome and Labitis. It likes to boil things down to two competing theories that can't both be right, and then devise highly artificial tests to see which one wins. But this methodology often involves lumping a lot of different factors together that may not all work with each other, and it disallows the possibility that both theories may contain elements of truth. System One and System Two are each collections of many different processes, which often work together in mutually supportive ways.
28. Gerald Matthews, Moshe Zeidner and Richard Roberts, *Emotional Intelligence: Science and Myth*, Bradford/MIT Press: Boston, MA, 2002; Robert Sternberg, *Practical Intelligence*, Cambridge University Press: Cambridge, 2000. Howard Gardner, op. cit.

ACKNOWLEDGEMENTS

I am truly grateful to all of the following for their generous help – of many different kinds – in the shaping and writing of this book: Miles Barker, Ian Barnes, Stephen Batchelor, Sarah-Jayne Blakemore, Valerie Bonnardel, Fritjof Capra, Malcolm Carr, Margaret Carr, Art Costa, Matthew Crawford, Jean Decety, Thomas Fuchs, Howard Gardner, Gabriel Gomez, Stephan Harding, Paul Howard-Jones, Jean Knox, Michael Leunig (sorry that Mr Curly didn't make the final cut), Terry Locke, James Pennebaker, David Perkins, David Perry, Rolf Pfeifer, Graham Price, Greta Raaen, Richard Sennett, Louise and Guy Thwaites, Tamar Tolcachier, Harald Traube, Maria Alessandra Umiltà and Michael West. Sometimes just a fleeting conversation, which they may not even remember, sowed a seed that germinated over time into a frond or a branch of this book. I am honoured to have known two titans in the fields of embodied cognition and systems thinking who are no longer with us: biologist Brian Goodwin (a colleague at Schumacher College in Devon) and Francisco Varela – the only person I have ever met who could talk European phenomenology, neuroimmunology and Tibetan Buddhism fluently, and at the same time. It will be obvious that I have been deeply swayed, in addition, by the pioneering work of researchers such as Lawrence Barsalou, Andy Clark, Hugo Critchley, Chris Frith, Arthur Glenberg, Mark Johnson, George Lakoff, Anil Seth and Timothy Wilson.

Much thinking about real-world intelligence was stimulated by discussions with my (now ex) colleagues at the Centre for Real-World Learning at Winchester: Janet Hanson, Ellen Spencer, Jenny Elmer, Roberto Webster and especially Bill Lucas. My friends Sebastian Bailey, Steve Brickell, George Chamier, Judith Nesbitt and Jonathan Rowson were all heroic in offering to read an earlier draft, despite their own backlogs and busy lives, and in providing sometimes painful, but always useful and perceptive, feedback and suggestions. Many thanks to Heather McCallum and her colleagues at Yale for taking the book on, and again for much good advice that has helped to improve the book enormously. And to one anonymous reviewer for gentle and penetrating comments that really helped. Judith Nesbitt gets a lot of points for allowing me to colonise the dining table for long stretches without (much) complaint, and generally cutting me a good deal of slack, and offering continual support, during the writing.

*****

INDEX

Note: Page numbers in italics refer to illustrations.

Ekman, Paul 108
elderly people and exercise 258–259, 260–261
electrical conductivity of skin 60, 134–135, 213
electrical impulses and whole-body communication 81, 83, 101
electroencephalograph (EEG) patterns and biofeedback 249–250
electronic communication *see* digital communication
eliminative materialism 139
Elizabeth II, Queen 185
Ellis, Rob 67
embarrassment as emotional mode 120
embodied cognition 2–3, 12, 41
and brain 80–81
embodied intelligence 9, 25, 138–166, 290–292
and anticipation 72–77, 180–182, 213–214
and peripersonal space 194–201
and physical expertise 223, 225
and body and movement 38–44
human hand 41–44, 95–96, 195–196
movement and thinking 164, 248, 257–262
coping with unfamiliar situations 65–66, 75, 76
craftiness and expertise 219–241, 284, 285–288
and creativity and imagination 163–164
and language 139–140, 147–155
and mathematics 159–162
methods for improving 248–262
and physical exercise 10, 273–274
and senses 66–72
and social structures 11, 266–292
and thinking 155–159, 228–231
understanding ideas 68–69
see also interconnection of body and brain

emotional intelligence (EI) 9, 104, 291
emotions and feelings 102–137
and abstractions 145–146
blended emotions 120–121
and boarding school education 272, 273
and body 106–107
and basic emotional modes 108–120
and cognition 130–137
effect of mood 131–133
experience and decision-making 134–137
reasoning and emotions 102–103, 133–134
and consciousness 184
control and inhibition of 123–127
and crying 128–129
and cultural difference 120
and digital communication 280
and family influence 121–122
'felt sense' and psychotherapy 255–257
and intelligence 104
purpose of 102–103, 104–108
and skin and touch 58, 61
and social resonance 212
and theories of the mind 26–27
and thinking 5–6, 131–132
empathy 125–126, 163
digital communication and reduction in 279–280
and social resonance 211, 212, 216, 217–218
endocrine system *see* hormonal system
enquiry as emotional mode 114–116
environment and augmentation of intelligence 193–218, 292
envy as emotional mode 121
Evans, Francis 202–203
evolution of cells 33–35
exercise *see* movement; physical exercise
expectations
and action 65